景观森林

景虎国际/龙赟 著

设计实践

2006
LANDHOO
2016

东南大学出版社·南京

景观森林 / 景虎设计实践

一、实践历程——路在景虎的脚下

景虎的故事是从2006年开始的。

公司刚成立时,不管多么不靠谱的甲方,只要有一点机会都得抓住啊!李晓岚、阎莹、李思远等最早的一批同事一起在时代天骄小区里埋头画图;第一次承接悉尼湾这样的百亩项目的喜悦;汶川地震后在温江继续战斗;铁面陈总滔滔江水般的怒言;福州的烈日下与母丹、雷冬一起爬了四年还没爬完的山顶公园;西安的雪地里种植物;融侨公园获奖;领导视察南江滨……这是我,景虎国际的总经理和设计总监,在公司成立十周年时的感言。

从最初在成都时代天骄小区里的拥挤办公室,到迁入写字楼,一群技术理想主义者,一群用作品说话的设计者,一群不断提升的攀越者,一群不畏艰难的乐天主义者不断围聚起来,一同走上景观设计"执着于景·敬天爱人"的不懈追求之路,始终秉承着"敬重自然、遵循天道、践行环保"的信念。我们切实探求景观设计、景观工程、苗木生产的产业一体化模式,并精专于景观研发、景观规划设计、景观产品集成、施工现场服务等项目管理与策划增值服务。

这一路是怎么过来的呢?这本记录景虎成长的书册不仅仅是纪念,更是小结,也是这个团体人人参与、不断总结的思想结晶。

2006年:萌芽。 关键词:时代天骄、一无所有、水慕清华、人员、期待正式项目

本着诚信务实的态度,终于接到水慕清华的项目。景虎对这个项目投入了大量精力,到现场与施工方一起种树,购买雕塑小品、硬景材料。所幸的是,这个项目完工后在当地引起广泛关注,并受到一致好评。后来还被收录在《2010年中国景观设计年鉴》中。

2007年:存在。 关键词:汇融项目、鸿阁一号、保利项目、搬家、设计师、团队组建

经过一段艰难的时光,公司慢慢创建了设计团队和研发团队。迎来了第一个一线开发商的项目:保利-洞庭东岸。鸿阁一号、汇融名城等项目终于以实景呈现。公司技术得以提升,设计流程也更加规范。

2008年:烙印。 关键词:地震、金融危机、研发、为大公司做设计

作为一个刚起步、还未站稳脚跟的新公司,我们经历了接二连三的冲击:地震、金融危机、众多项目因开发商的土地转让而流产……但景虎坚持同舟共济,降薪不裁员。在这一年里,研发部对地产景观的系统研究也从未中断,使理论研究工作持续推进。

2009年:坚守。 关键词:融侨、论坛

这一年是景虎大发展的一年,也是最艰苦的一年。景虎开始了和融侨集团的紧密合作。融侨集团的设计总监是施工单位公认的"陈杀手",他的严格要求几乎压得我们喘不过气,施工图和种植设计图经过二十几轮修改才定稿。我们在城市之间不停穿梭……年终的时候,我们终于舒了一口气,熬过来了!这一年,我们还成功举办了"2009地产景观高峰论坛"。

2010年:涅槃。 关键词:上海、经历、教训、经验、长三角系统了解、市场

2010年,景虎又搬家了。办公环境越来越好,各方面都步入正轨。但这一年也得到了不少的教训:在企业扩张方面,上海分公司组建又撤回的经历让我们对市场有了更清醒的认识,尤其是对长三角城市的设计市场有了更深刻的了解。研发部经过三年努力研发而成的《迈向产品系》也逐渐成形。这一年,景虎开始涉足苗圃业,为下一个五年计划,即产业链一体化做好准备。

2011年:甲方大爷。 关键词:幸福感、Hold住、视野、奖金制度

面对依然强势的客户,我们无奈地称他们为"甲方大爷"。我们期待将设计行业的买方市场转变为共赢市场,用我们的

实力证明设计产品的质量，也许到那时甲方便不再是"大爷"了。同时，这一年是景虎着力提升幸福感的一年，公司实现了大幅度加薪，并建立了客观公正的奖金分配制度，激发了员工的潜力。

2012年：《迈向产品系》。关键词：研发、沟通与交流、团队建设

《迈向产品系——地产景观之路》这一辛苦孕育的研发成果是景虎研发部对当时地产景观风格的详细梳理，这本书出版后在业内广受好评，乃至成为了开发商的工具书！这一年，景虎的成长也有赖于高素质团队的建设。与此同时，景虎提倡通过"提升技术、改进工作方法、提高工作效率、建设数据库"等手段对员工的工作方式进行改进，大大改变了设计师的亚健康状态。

2013年：提升。关键词：人才建设、团队智慧、发展计划

这一年是景虎持续成长和提升的一年。人才的全方位建设、注重个体的培训和训练，使个人发展和公司发展同步提升。

2014年：待续。关键词：平台——行业平台、事业平台、个人发展平台

大事件一：景虎着手搭建文化传媒平台，最终形成"设计—传媒"一体化的产业发展方向。"DE地意文化"对外建立景观行业最具影响力的企业精英组织——DE私董会，对内则作为景虎团队团结的纽带。"DE地意文化"的活动均以"助力景观设计＋，服务园林产业链"为宗旨。

大事件二：2014年年底，景虎组建文旅团队，从此开启了璞旅、田园、文创等文旅细分领域的探索之路。

2015年：应变。关键词：整合团队、寻求突破

在2015这个"不寻常的变化之年"，景虎也在"阵痛"中成长。一方面，设计（人居、文旅）团队、传媒团队等各个部门进行了人力资源的优化整合；另一方面，"在这个最坏的时代，也是最好的时代"，景虎各个团队也在摸索中前进，三思而后行：人居力争资源、文旅构建资源、传媒整合资源。无论是项目投标，还是项目开发，2015，景虎都在市场的淬炼中无畏前行。

就这样，从一个人的理想到一个团队的理想，景虎国际坚持强调对景观的全面认识、理解和创造，坚持不懈地从文化背景、市场需求、区域条件和可持续发展的角度来审视景观行业，深入理解技术发展、市场变化带来的影响，以及其与经济、文化现实的矛盾，积极创造景观的新语言和新文化；在研发方面致力于在理论与市场的平衡中研发产品，加强与国内外相关学术机构的沟通、交流与合作；在标准化方面为开发商提供景观产品系研发服务，解决客户在快速开发过程中遇到的质量及成本控制等问题，通过构建标准化设计达到对项目的有效管控。

由此，景虎国际不断发掘特定条件下景观与市场、景观与社会构成中潜在的景观价值和社会价值，为不同文化生活、不同地区环境下的景观与规划塑造出具有标志性和导向性的作品。在成都、重庆、福州、西安、南京、合肥等中国中纬度区域的城市中，不断涌现出景虎国际的标志性作品。这些成果正是景虎国际在设计实践中的一个个坚实脚印。

二、设计实践——从思考到践行

实践的景虎，行进的景虎——这个团队有理想、有研究、有创新，在对话、在行动、在传播……景虎人是幸福的，景虎人正做着幸福的事。

景虎在思考

景观是人与人、人与自然的关系在大地上的烙印，是自然与美印记在人心中的永恒结合体。景观规划设计何尝不是对自然的关怀，对文化的关怀，对生存和生活的关怀。因此，唯有人类本着仁爱的态度与自然相互作用，才能取得生动的和谐，平衡的大美。

景虎国际通过不断地学习、思考和创新，谨慎而大胆地进行每一次设计探索：与建筑学、城乡规划学、环境艺术、市政工程设计等学科和行业紧密联系，同时掌握好自然系统和社会系统的多方面知识，协调人与自然、人与城市的关系。道法自然，仁爱万物，让现代生态伦理与景观设计相互交融。力求设计出充满个性创意与人文气质的作品，同时倡导不盲目追求物质享受、多为社会奉献的绿色生活理念。

本书正是如此深入思考的作品呈现。景虎团队在深刻理解项目场所精神的基础上，综合协调各方面因素进行创作，包括文化气质、环境特性、功能需求、资金投入量等。

景虎在发声

"我一开始就没有把景虎国际简单地视为一个求生存发展的公司，如果自己的思考、研究、推动是有益的，为什么不对景观设计业界的一些困惑和流弊发出正当的回应？对社会的担当、对业界的影响，我义不容辞。"

部分文章摘录：

"我们应该思考如何将传统的居住模式与现代功能结合起来。在目前的阶段过去之后，国民的审美将发生很大的变化，会真正重视和尊重自己的文化、自己的生活方式……商业化是目前中国居住区存在最大的问题的症结所在。"

——《在中国过别人的日子》，龙赟（《景观设计学》第25期）

"景观已经沦为地产交易过程中的赠品！甲方大爷和我们都得思考：谁在为景观设计买单？"

——《谁在为景观设计买单》，龙赟（《国际新景观》第38期）

"2010年第二届成都景观论坛上，我们在呼吁：警惕新'蜀山兀，阿房出'；看着路边哭泣的大树，看着村头、山头、田头日渐消失的大树，不由想到'开发商要流着道德的血液'之语，我想这个道德，不光是社会道德，也包含着生态伦理与生态道德。"

——《血泪花园的联想》，龙赟（《国际新景观》第39期）

"东方，被国际化与现代主义摧毁了的东方，仅仅是商品，不再是生活的载体……西方更像是马赛克，不管深色、浅色还是杂色，我们总能从中看出整体的色彩与图像脉络。时代在色块中演进，生活在演进中还原。身临西方城市，你才觉得中国对于历史文化名城、传统街区的保护有多么可笑。"

——《东西方》，龙赟（《国际新景观》第41期）

"在公共景观设计领域，学术界也常常点起行政体制内的'山火'。概念层出不穷（山水园林城市、田园城市、绿道、城市风貌、新农村，等等），每一个概念都伴随着一轮设计上的鼓吹，拆了建、建了拆……我们的教育需要告诉大家：真正做负责任的设计，真正做社会所需要的设计！"

——《景观理想的现实比照》，龙赟（《国际新景观》第42期）

景虎在研发

景虎拥有四大市场领域：地产景观、文旅景观、公共景观、生态景观。景虎人从不把"创新"简单地理解为"与众不同的形式"，而是赋予每一处景观不同的场所精神和生命特征。在这个过程中，景虎始终致力于对设计前沿问题的研究，在地产景观设计、公共景观设计、文旅景观设计等多方面形成了主题化、本土化、生态化、情境化四位一体的独到技术和独特趋势。

通过于2007~2010年间对全国21家优秀开发商的676个楼盘景观进行系统考察和研究，景虎收集了大量的第一手资料。在剖析景观产品系意义的同时，探究其研发体系、实施体系以及景观功能模块等多个体系的内部环节，从而为企业构建景观产品系提供参考框架。其研究成果《迈向产品系——地产景观之路》已于2012年3月由上海科学技术出版社出版发行，受到广泛关注与好评，被作为行业手册。其后，景虎国际与复地集团、长虹集团、金融街集团合作完成了其景观标准化手册的研发任务。

景虎研发部关于景观产品系定制（设计）的研究成果还包括《景观产品系指引手册》《景观设计管理手册》与《景观设计标准及标准图集》等。除了在地产景观方面的钻研外，景虎也进行商业景观、文化旅游景观和生态景观领域的研究，并努力将其从研究转化为实践。

景虎国际一直希望做一个引领者，而不甘为跟随者。景虎国际将持续加大对研发的投入力度和对设计的深入研究，将产品系研发作为创新设计的助力器，并为项目研究提供理论支持。

景虎在对话

只有通过踏踏实实的交流和学习，才会具备更为广阔的视野。这些年来，景虎国际多次主办设计类的学术沙龙，各界朋友、同行、兴趣者等闻声而来。"浣花荟"系列学术沙龙曾邀请法国国立尼姆美术学院、北京大学景观设计研究院的安建国老师担任主讲嘉宾，举行"法国城市景观之路"讲座；上海资深景观设计师俞昌斌先生来讲解他的《源于中国现代景观设计：空间营造》，探讨如何寻求传统内涵与现代设计的平衡点，寻求传承中国文化传统内涵的现代表达方式。景虎国际还策划了"我的景观十年"设计随想类活动，活动参与方包括学术界、教育界人士，以及知名开发商和景观设计公司等，大家就案例分析、设计灵感之源、设计过程中的点滴等畅所欲言，分享了各自成功与失败的经验，引发了更有力量的设计思潮。

景虎人还经常参加国内举办的各种论坛、展会、培训、交流等。如北京大学建筑与景观设计学院"弹性城市"论坛、哈佛设计研究生院兼职教授克里斯·里德的演讲、主题为"景观设计与社区营造"的上海万科论坛等，也曾在第三届成都景观论坛，就"景观创新与行业发展""研究·设计·产品"等议题和与会同行进行了深入的研讨。此外，景虎在2012~2014年间连续三届担任全国高校景观设计优秀毕业作品展评委，并参与高校景观设计优秀毕业作品交流会暨高校学生论坛。

"学习、沟通、交流"逐渐成为公司的常态。通过这一系列行业内的、跨行业的、与国际接轨的、储备新生力量的对话和交流活动，景虎国际逐步展现出更强的创新力、行动力和执行力。

不仅仅是文化传播

自2014年起景虎国际加大了行业平台的建设力度。2014年2月开始运营"DE景观广角"微信公众号，每天发布景观行业的新资讯：世界前卫的创意设计；人居、文旅、市政等各类景观赏析与评论；景观工程技术推荐与博览；景观行业的培训和教育；著名企业和特色推广；景观的苗木商情……

景虎发起的DE景观私董会涵盖全国300余家主流设计企业，为景观设计企业的管理者们提供了交流互动的平台。至今已在成都、重庆、西安、北京、广州等各大城市举行了见面交流会，形式活泼生动，内容丰富精彩，使得景虎自身和参与的会员们都受益匪浅。

同时，景虎国际还开展各种形式的专业培训、行业考察等活动。我们在做的，不仅仅是一种文化传播、话语传递；景虎团队旨在成为有社会责任感和行业担当的观察者和引领者。

三、企业文化

"要时刻想想客户还能信任我们多久。"正是抱有这样的诚恳态度，景虎国际团体中的每一个人都尽力地回报合作方的信任和托付。我们"从心出发"的理念赢得了越来越多包括国际性企业在内的大公司的合作意愿。

紧紧围绕"For Land, For You"的核心价值观，景虎国际在项目实施中不断克服并跨越景观设计公司常见的问题，例如异地交流、效果把控、设计创新、精准实施等。2008年起，景虎采用"协同设计"的工作方式：按"标准作业流程"，达到所有工种的全流程合作，以保证项目最终的可实施性。这其中包括对外合作和内部系统合作两个部分。对外合作包括与同行设计单位、专业机构、开发商等的合作，通过与各合作单位深入沟通，对项目定位、设计细节等进行深入探讨，为管理者提供良好的决策背景。在内部系统合作中，项目团队由各领域专业人员组成，共同参与设计方案制订、施工图绘制，以及后期服务等各个阶段，在实践中让多方受益，也让合作更顺畅，工作更有效率。

员工价值与幸福密码

所思、所做的这一切，都来自景虎国际这个生命有机体，来自这个亲密的大家庭式学习型组织。"敬天爱人"，景虎为团队中的每一个人都提供良好的成长环境和支持动力，让每一位员工在这里都能够愉快地工作与学习，从而使大家在合作中成长，在成长中更加紧密地合作、共同创造。在这一过程中，个体的价值、工作的意义融为一体，如同获得了春风雨润的田地，长出宝贵的资产。

景虎一直致力于员工幸福感的营造：让员工深感努力被认同、成绩被认可；建立完备的培训和发展体系，通过个人成长计划和导师制度，使每位员工都在岗位上有所作为，为每位员工规划出清晰的职业发展道路；在薪酬福利方面，景虎提供了富有竞争力的薪酬和完善的福利政策，建立了公平的分配体系；在对员工进行及时激励的同时，也建立了员工问题反馈的畅通渠道。

做一个幸福的人，才能做成幸福的事。

积极、简单、透明

公司的每一份成绩都是大家的荣誉，景虎国际以包容的心态破除思维定式，用整体而客观的历史发展观看待各种景观形式和创作手法，随时做记录、查资料、梳理文献、整理研究笔记等。

在心思意念、团队合作、人际关系、工作程序，以及劳动分配等各个方面，我们都提倡并注重简单化，令员工能够在轻松、简单、纯净的环境中安心工作，专注于设计创作，而不为其他所影响。

运作透明是与公平、公义和公正联系在一起的。让每一位员工都能成为主角，主动参与经营，减轻层级官僚气息。在经营上采用"阿米巴"模式，建设多个拥有明确志向和目标、持续自主成长的独立组织。以各个"阿米巴"的领导为核心，让团队自行制订各自的发展计划，并依靠全体成员的智慧和努力共同达到目标，进而实现"全员参与经营"的最终目的。公司的绩效指标直接落实到团队的项目组上，从而减少冗员，有效地应对环境变化，提高每个人的能动性。由此培养出一批又一批的实战型人才，从而让公司整体运作更加有效，同时也有力地支持了协同技术管理。

四、未来的路

回首景虎这些年来走过的道路，虽不够完美，但也没留下太多遗憾。景虎团队始终在努力，坚持着"工作是美丽的"这一理念。我们将工作中出现的问题看作自己成长的机遇，敢于面对和超越，每一次挑战都会带来多一点的进步。随着中国建设行业"白银时代"的到来，以及众多国际景观公司的入驻，本土的景观设计公司面临着更大的压力。相信压力也就是动力，总能有智慧和力量开出一条前行的路，且将会越来越宽阔。

龙 赟

2016年4月

于成都

LANDSCAPE FORESTS: PRACTICE OF LANDHOO DESIGN

I LANDHOO'S TEN-YEAR HISTORY OF PRACTICE

The story of LANDHOO began from 2006, "When the firm was first founded, we seized every bit of opportunity regardless of how ridiculous the clients were. Xiaolan Li, Ying Yan, Siyuan Li and other colleagues buckled down to work in our informal office in a residential area; the excitement when we first obtained a big project as Sydney Bay; the days of keeping working in Wenjiang after the earthquake; the raging face of Manager Chen; climbing the mountain-peak park that we had tried and failed for four years with Dan Mu and Dong Lei; planting in Xi'an on a snowy day; winning the award for the Rongqiao Park; government leaders' visit to the Nanjiangbin project… "

As the general manager and chief designer of LANDHOO, I present the above at the tenth anniversary of the company's foundation.

From the crowded, informal office in Shidaitianjiao Residential Area of Chengdu, to the magnificent office tower, we gathered as a group of idealists, a group of designers who voiced through work, a group of climbers who sought to higher goals, and a group of optimists who despised difficulties. We keep the faith of "ardent love for landscape and great respect for nature and human." We stay loyal to respecting the nature, following the natural laws and advocating environmental protection. We explore the integration mode of landscape design, construction and nursery production. We devote to landscape research and development, planning, product integration, as well as project management and value-added services such as on-site service.

How do we summarize the stories of LANDHOO in the past ten years? This publication documents the growth of LANDHOO. It is not only about memoirs but also a milestone, a collective achievement of all the team members.

2006: Sprout. Keywords: Shidaitianjiao Residential Area; Zero-start; Shuimu Qinghua Project; Team Members; Desire for Formal Project

Thanks to our honest and practical attitude, we finally won the Shuimu Qinghua Project. We spent great amounts of effort on the project, including planting trees with the contractor, purchasing ornamental structures and hardscape materials. Fortunately, the project ended well and gained wide attraction and good reputation. It was included in China's Landscape Design Yearbook of 2010.

2007: Existence. Keywords: Huirong Project; Hongge No.1 Project; Poly-Dongting East Bank Project; Relocation; Designers; Team Building

After a tough time, LANDHOO established the design and R&D teams. We also won a project developed by a premier developer — Poly-Donting East Bank Project — for the first time. Other projects including Hongge No.1 and Huirong Mingcheng came to reality. Our technology was improved, our design process became more standard.

2008: Imprint. Keywords: Earthquake; Financial Crisis; R&D; Collaboration with Major Companies

As a start-up and unsteady company, LANDHOO experienced many blows: earthquake, financial crisis, and abortion of some projects due to land transfer by the developers… Even under such circumstances, LANDHOO kept solidarity by cutting down salaries instead of cutting employees. In this year, the crew of R&D department kept on their systematic study over real estate landscape.

2009: Holding on. Keywords: Rongqiao Group; Forum

2009 was the year of major development for LANDHOO and also the hardest year to the team. We began close cooperation with Rongqiao Group. Mr. Chen, the design director of Rongqiao Group, was called "the killer" by the local construction companies. His strict requirements weighed us down. Usually it took more than 20 rounds of revision to finalize the construction documents and planting design. In this year we had to travel among cities… By the end of the year, we finally took a breath. We survived! We also succeeded in holding the 2009 Real Estate Landscape Summit Forum.

2010: Nirvana. Keywords: Shanghai; Experiences; Lessons; Profound Understanding on Yangtze River Delta; Market

LANDHOO relocated again. The office was better and everything was on track. But we also learned a lot of lessons: as to the corporate expansion, the establishment and following dismissal of Shanghai branch made us more profoundly aware of the market, especially the design market in Yangtze River Delta. After three years of endeavor, the *Towards Product Lines* by LANDHOO R&D department began to take shape. In this year, we took our first step in nursery industry, preparing for the integration of full industry chain which was recognized as the next five-year goal of LANDHOO.

2011: Client the boss. Keywords: Happiness; Holding on; Vision; Bonus Scheme

Facing the aggressive clients, we reluctantly called them "Client the boss". We were looking forward to the shift from the buyers' market to a win-win market, where we could prove the quality of our work and where the clients stopped being the boss. Meanwhile, LANDHOO focused on improving the employees' sense of happiness by not only realizing a sharp increase of salary but also establishing a fair and equitable bonus scheme that encouraged the team members.

2012: *Towards Product Lines*. Keywords: R&D; Communication; Team Building

Towards Product Lines — Real Estate Landscape Product Lines by LANDHOO R&D department represented thorough organization of the contemporary real estate landscape style and had become a reference book for developers. All through the year, the growth of LANDHOO was largely contributed by establishing a high-quality team. Meanwhile, employees' working methods were improved through better technology and working approaches, higher efficiency and database expansion. The health status of the designers was significantly improved.

2013: Improvement. Keywords: Talent Pool; Collective Wisdom; Plan for Future Development

2013 was a big year of growth and improvement for LANDHOO. We focused on establishment of comprehensive human resource system and talent pool, as well as individual training that was conducive to the growth of both the employees and the company.

2014: Continuance. Keywords: Platform — Industrial Platform; Career Platform; Individual Development Platform

Big Event One: LANDHOO started to build a culture media platform, DE Culture, with an ambition of forming the integration of "Design — Media" corporate industrial development. Externally, DE Culture founded the most influential business elite organization in landscape industry of China — DE CEO Group; internally, it bonded the LANDHOO's team. All the events hosted by DE Culture during the year were with the same goal of "facilitating landscape design+, serving the landscape industry chain".

Big Event Two: At the end of 2014, LANDHOO established its own team for cultural tourism business, and thereby started its exploration in the sub-fields such as tour of simplicity, rurality, and cultural creation.

2015: Flexibility. Keywords: Team Consolidation; Seek for Breakthroughs

In 2015, the year of unusual changing, LANDHOO grew in the "labor pain". On one hand, departments such as the design (residential design and cultural tourism) and media teams achieved optimization and integration; on the other hand, in this both "best of times" and "worst of times", each team of LANDHOO kept groping prudently. The residential design team endeavored for resources, while cultural tourism team structured resources and media team integrated resources. In 2015, LANDHOO kept moving fearlessly, whether in project bidding or the design process.

From one man's dream to a team's dream, LANDHOO insists on comprehensive understanding and creation of landscape, and examining the landscape industry through the lens of cultural context, market requirement, regional conditions and sustainability, thoroughly understanding the influence on technology development and market changes, the conflicts with economy and culture, as well as creating new language and culture in landscape. As to R&D, we focus on developing products, strengthening the communication and cooperation with both domestic and overseas academic institutions. Regarding the aspect of product standardization, LANDHOO provides the developers with R&D services of landscape product lines to balance the quality and cost control issue in the process of rapid development and achieve effective control over the project through standardized design.

Therefore, LANDHOO keeps exploring the potential landscape and social values between landscape design and market, as well as between landscape and social composition, providing symbolic and directive work of landscape design and planning for different cultural lifestyles and regional circumstances. LANDHOO's practice flushes in Central-China cities, such as Chengdu, Chongqing, Fuzhou, Xi'an, Nanjing, and Hefei. The works themselves speak the philosophy of LANDHOO.

II PRACTICE — FROM THINKING TO INITIATION

LANDHOO in practice, LANDHOO in process—LANDHOO is a team with ambition, study and innovation. LANDHOO is a platform for dialogue, action and communication... LANDHOO staffs are happy; LANDHOO staffs do cheerful things.

LANDHOO is Thinking

LANDHOO believes landscape is the imprint of relationship between human and human, and between human and nature. Landscape is the eternal home and branded on men's heart. Landscape planning and design is the care for nature, for culture, and for subsistence and life. Therefore, the dynamic harmony and balanced beauty can only be achieved through respectful touch on nature.

LANDHOO never stops learning, thinking and innovating. Every exploration on design is conducted in discretion and dauntlessness: we work closely with related disciplines and industries including architecture, urban and rural planning, environmental design, civil engineering, etc.. Meanwhile, we grasp the knowledge of natural and social systems, and harmonize the relationships between human and nature and between human and city. We mimic the nature, care for all creatures and advocate integration of modern ecological ethic with landscape design. LANDHOO's work strengthens both creativity and cultural quality, and also highlights the life philosophy of avoiding blind pursuit of material comforts and the spirit of devotion.

All LANDHOO's practices in this collection are based on thorough consideration, deep understanding on the spirit of the place, with integration of various elements, such as cultural features, environmental characteristics, functional demands, and financial investment.

LANDHOO is Voicing

I never regard LANDHOO as a company only seeking for survival. Why do not we give proper response to confusions and popular yet improper practices in the landscape design industry if our own thoughts and studies are beneficial? I am obligated to the social responsibilities and industrial influences.

Excerpts of Publications by LANDHOO:

"We should think of how to integrate traditional living modes with modern functions. The national aesthetic will change dramatically when the current trend is over. We will value and respect our culture and lifestyle... commercialization is the major crux of the design of residential areas in China."
— Living Foreign Life in China, Yun Long (Landscape Architecture Frontiers, No. 25)

"Landscape has fallen to become the gift during the real estate selling. Clients and designers should think about a question: who is paying for the landscape design?"
— Who is Paying for the Landscape Design, Yun Long (International New Landscape, No. 38)

"On the 2nd Chengdu Landscape Forum in 2010, we pointed out that we should be on guard against the emerging phenomenon of 'Epang Palace was built as all woods of Shu Mountain were stripped'. Looking at the miserable trees in cities and disappearing trees in the villages, hills and fields, we cannot help recalling the saying 'developers should have moral blood'. I think the moral code here does not only refer to social ethics but also ecological ethics."
— Thoughts on Garden of Blood and Tear, Yun Long (International New Landscape, No. 39)

"Orient, where has been being corroded by internationalization and modernism, is nothing but a commodity that is no longer suitable for living... Occident is more like mosaic. Whether it is dark, light or mixed, we can always figure out the holistic color and image. Eras evolve in these colors while life changing and reintroducing; then you start to realize how ridiculous China's protection over historic cities and traditional neighborhoods is when you visit and understand western cities."
— Orient and Occident, Yun Long (International New Landscape, No. 41)

"In the realm of public landscape design, academic experts often help governments hyping up movements. Coming along with different concepts (garden city, greenway, urban appearance, new countryside, etc..), the direction of design changes, resulting in new round of urban construction and demolition... The landscape education should teach us to be responsible to our design, and meet the real society needs."
— Landscape Ideal and Reality, Yun Long (International New Landscape, No. 42)

LANDHOO is Researching and Developing

The business of LANDHOO covers four major areas: residential landscape design, cultural tourism landscape design, public landscape design, and ecological landscape design. We never simply interpret innovation as to present an unprecedented style; instead, every project should have its own unique sense of place and life characteristics. During each design process, LANDHOO has long devoted itself to the study on the issues of design frontiers, and the development of its original technique and prediction in theme, locality, ecology, and scenario, ranging from residential landscape design, public landscape design, and cultural tourism landscape design.

Primary materials were collected during the systematic survey and research of 676 residential landscape design projects in China by 21 Chinese major developers from 2007 to 2010. We analyzed the internal link among diverse systems including R&D, construction, and landscape functional modules etc., while exploring the meaning of the landscape product lines in order to provide a general frame of landscape products for developers. We documented and published the research findings and results in a book, titled *Towards Product Lines — Road to Real Estate Landscape*, by Shanghai Science and Technology Press in March 2012. The publication has attracted wide attention and raised good reputation as a guidebook. After that, LANDHOO collaborated with Forte Group, Changhong Group and Financial Street Holdings in developing their own landscape standardization.

The research achievements of LANDHOO's R&D department also include *the Guidebook of Landscape Product Lines, Landscape Design Management Manual* and *Landscape Design Standard and Standard Drawings*, etc. Besides the in-depth research in residential landscape design, LANDHOO also conducts researches in the areas of commercial landscape design, cultural tourism landscape design, and ecological landscape, and attempt to apply the research results into practice.

LANDHOO has long intended to be a leading group of research. We will continue to increase the investment to R&D and study on landscape design, taking our products as the catalyst to innovative design and providing theoretical supports to project research.

LANDHOO is Communicating

I like to communicate and learn earnestly to broaden my horizon. In the past years, LANDHOO has hosted a series of design salons. "Huanhuahui" serial academic salons used to invite Jianguo An, visiting lecturer at Ecole Superieure des Beaux-arts de Nimes and the Graduate School of Landscape Architecture of Peking University, to give his speech on "Development of French urban landscape". Changbin Yu, senior landscape architect from Shanghai, shared his work, Modern Landscape Design Originated from China: Space Creation, to discuss how to balance the traditional meaning and modern design and seek for modern expression of the traditional Chinese culture. LANDHOO also held the serial activities "My ten years of landscape design". Participants ranged from academic and educational experts, to renowned developers and landscape practitioners. Case studies and stories of design inspiration and process were discussed on such activities to evoke and inspire broader thoughts.

My colleagues and I often attend domestic communication events. For example, we attended the Forum on Resilient City held by the College of Architecture and Landscape Architecture of Peking University, the lecture given by Chris Reed, visiting professor of Harvard Graduate School of Design, the Shanghai Vanke Forum on Landscape Architecture and Community Establishment. We were also invited to the 3rd Chengdu Landscape Forum and shared our views on the topics including "landscape innovation and industrial development" "research, design, and products", etc.. In addition, I was invited as a jury member of the Chinese Landscape Architecture Graduate Works Exhibition from 2012 to 2014, and participated in the Outstanding Graduate Works Communication and Student Forums.

"Learning, communicating, and exchanging" is becoming a normalcy of LANDHOO. I delightfully witness the innovation, action and execution in LANDHOO through these cross- and inter-industry international dialogues and communications participated by emerging talents and young designers.

More than Cultural Communication

LANDHOO has put more efforts in establishing the industrial platform since 2014. The WeChat public ID "DE Landscape Panorama" was formally initiated in February 2014. It releases industrial news every day, including reviews and comments on recent projects of residential landscape design, cultural tourism landscape design, public landscape design; landscape engineering technology; industrial training and education; famous companies and promotion updates; nursery market information, etc..

The DE CEO Group initiated by LANDHOO now covers more than 300 major design companies, providing an interactive platform for company managers. Serial meetings have been held in Chengdu, Chongqing, Xi'an, Beijing, and Guangzhou. Through various forms and agendas, while stimulating larger communication, the members of DE CEO Group bond.

Meanwhile, LANDHOO will conduct various professional trainings and industrial visits. What we are doing is not only cultural communication; instead, my team and I aim to become observer and leader with social responsibilities and industry missions.

III Corporate Culture

What I often stress is "always remind to ask ourselves that how long the clients trust us". With such attitude, everyone in LANDHOO is trying to pay back the clients' trust. Therefore we earn collaboration opportunities from more and more major companies, including international enterprises.

Based on the core philosophy of "For Land, For You", LANDHOO overcomes and transcends beyond the common issues encountered by landscape architecture companies, such as long-distance communication, effect control, design innovation, accurate and refined implementation, etc. We have adopted the "synchronous design" method since 2008. Following "standard process" — synchronous landscape design, all professionals collaborate with each other to guarantee the final implementation of projects. There external and inner cooperation are included. The external cooperation includes collaborations with other design companies, institutes and developers in order to better position the project, refine the design details and provide solid background for decision-making. In inner cooperation, each project team is composed of professionals of various backgrounds who work together in the entire design process, construction document producing, and other further services in order to benefit the most, smooth the cooperation and enhance efficiency.

The Value and Happiness Code of Staff

What we believe and do is from the organic system of LANDHOO, from this close, big family and learning organization. as the motto "great respect for nature and human", LANDHOO provides a good growing environment for every team member so that everyone can learn and work happily here. It enables us to learn how to grow in cooperation. The value of individuals and the meaning of work are emphasized in LANDHOO's corporate culture. Everyone is benefited and rewarded.

LANDHOO has long been focused on the creation of staff's happiness: to make the employees feel recognized; to establish comprehensive training and development system; to create the sense of accomplishment for everyone through individual growing plan and mentor system; to develop a clear career plan for every employee. As to the salary, LANDHOO adopts a competitive salary system and welfares. We establish a fair bonus allotment system and a problem feedback system along with reward mechanism.

Yes. Only a happy man can fulfill happy business.

Vigorous, Simple, Transparent

Every achievement of LANDHOO is credited to the whole team. LANDHOO breaks the old mindset with embracing attitude. We consider all landscape styles and design approaches from a holistic and objective perspective through a lens of historical development. We record and document, read up literatures, and organize the materials and research notes in daily work.

We promote simplification in perspectives of mind, team work, relations, work procedure and reward system, to make all employees work in an easy, simple, and pure environment where they can focus on their design creation and avoid being affected by other unnecessary matters.

Transparent operation brings fairness, righteousness, and equity. We encourage every employee's participation in corporate operation, avoiding bureaucracy. Amoeba model is adopted in corporate operation to establish multiple independent organizations that have clear ambitions and goals and would help their individual development. Each Amoeba team develops their own plans and realizes the goals with the collective wisdom and efforts, thus LANDHOO realizes its ultimate goal of "full-participation in operation". The company's performance index is assigned to each Amoeba team in order to reduce supernumerary, be resilient to the market changes and stimulate individuals' potential. In this way, more and more talents will emerge that helps financial operation and supports the management of synchronous technology.

IV THE FUTURE

Looking backwards, the way of LANDHOO was not perfect but I do not have much regret. Because the whole team and I never stop endeavoring, and we believe that working is beautiful. We deem problems in work as opportunities for our growth. We stay fearless and determine to transcend. Every challenge brings us more improvement. As the domestic construction industry steps into the "Sliver Era" and more international landscape companies expanding their business in China, local companies including LANDHOO are facing greater pressure. However, I believe that the pressure is also impetus. We will have a bright future through our wisdom and power.

Yun Long

April 2016

Chengdu

目录
Contents

文旅
CULTURAL TOURISM

018　FUZHOU HUAI'AN GUESTHOUSE
　　福州淮安国宾馆

028　FUZHOU HUAQIAO PARK
　　福州华侨公园

036　ANATARA RESORT IN MOUNT EMEI, SICHUAN
　　峨眉山蓝光安娜塔拉度假酒店

040　WESTERN CULTURAL NEW TOWN IN CHENGDU
　　成都西部文化城

044　THE PUJIN HOTEL, CHENGDU
　　成都璞锦酒店

046　FUZHOU RONGQIAO WATERTOWN HOTEL
　　福州融侨水乡酒店

048　FIVE ELEMENTS FAIRYLAND VALLEY RESORT & SPA IN QINGCHENG COUNTY, CHENGDU
　　成都青城郡五行仙谷度假温泉

052　STARTING-POINT TOWN OF THE SILK ROAD IN XI-XIAN NEW DISTRICT, SHAANXI
　　陕西省西咸新区丝路原点小镇

056　DALY TIANYU LAKE INTERNATIONAL COMMUNITY
　　达利·天屿湖休闲国际社区

公共
PUBLIC SPACE

062　CHONGQING RONGQIAO PARK
　　重庆融侨公园

074　FUZHOU HUAI'AN RING ROAD PARK
　　福州淮安环岛路公园

080　FUZHOU NANJIANGBIN PARK
　　福州南江滨公园

088　HUAI'AN MOUNTAIN PARK IN FUZHOU
　　福州淮安山体公园

092　CHENGDU NEW WORLD RIVERFRONT PARK
　　成都河畔新世界滨河公园

人居
RESIDENTIAL LANDSCAPE

100 RONGQIAO WATERFRONT NEIGHBORHOOD IN FUZHOU
福州融侨外滩

114 KAMFEI RIVERFRONT NEIGHBORHOOD IN CHONGQING
重庆金辉春上南滨·江城著

118 CHENGDU HUAYUN TIANFU
成都华韵天府·三径

128 NANCHONG SHUIMU QINGHUA COURTYARD HOUSES
南充水慕清华

132 CHANG HONG NO.1 MANSION IN NANSHAN, MIANYANG
绵阳长虹南山一号

136 ZHIXIN EMPEROR STAR NEIGHBORHOOD IN FURONG ANCIENT TOWN AREA,CHENGDU
成都置信芙蓉古城·紫微园

140 QINGCHENG VILLA IN CHENGDU
成都青城郡

148 RONGQIAO MANSION IN NANJING
南京融侨观邸

156 CHONGZHOU YANVA SHANGLIN XIJIANG VILLAS
崇州炎华置信·上林西江

160 XI'AN TIANLANG WEILAN HUACHENG
西安天朗蔚蓝花城

164 JINCHENG OF TIANLANG DAXING RESIDENTIAL AREA IN XI'AN
西安天朗·大兴郡之锦城

168 LU TOWN DEMONSTRATION AREA OF THE WANHUA LUSHAN INTERNATIONAL NEIGHBORHOOD, CHENGDU
成都万华·麓山国际社区之麓镇样板区

174 MIANYANG CHANGHONG INTERNATIONAL NEIGHBORHOOD
绵阳长虹国际城

178 HENENG SEASONS NEIGHBORHOOD IN CHENGDU
成都合能·四季城

182 HENENG TREASURE AMBER NEIGHBORHOOD IN CHENGDU
成都合能·珍宝琥珀

184 CHENGDU HUIJIN COMPLEX
成都汇锦城

188 YITAI TIANJIAO NEIGHBORHOOD IN CHENGDU
成都伊泰天骄

文旅　CULTURAL TOURISM

福州淮安国宾馆
FUZHOU HUAI'AN GUESTHOUSE

1. 分期平面图
2. 隐秘在山林中的心灵归所

1. Phasing plan
2. A hotel hidden in mountain and forest as a place for taking refuge

　　淮安国宾馆位于福州南台岛最北端，紧邻淮安环岛路公园南侧。闽江蜿蜒至南台岛时被一分为二，南港称为乌龙江，北港仍称为闽江。从卫星航拍图上看，国宾馆恰好处于"龙头"位置，是一处名副其实的风水宝地。

　　特殊的地理位置赋予淮安国宾馆山林别样的细致与青翠，场地周边郁树繁茂、鸟鸣啾啾。为了塑造国宾馆浓厚的历史文化底蕴，设计团队在设计之初对场地山林进行实地踏勘，保留了一大批古树名木，为之后国宾馆环境的营造创造了极佳的植被条件。另一方面，场地中历史遗迹众多，如古县衙、五帝庙、临水宫、古窑址以及明朝状元翁正春、南宋理学家朱熹的遗迹等都得到了妥善保留，并且在后续的建设中进行有效保护与修复。这些丰厚的人文和自然资源为国宾馆的文化氛围营造创造了有利条件。

　　依托原始地形而建的休闲步道与疏林草地相结合的山地景观共同构成了共享空间，在这里，你可以观江景、品茶香；漫步山林，沐浴阳光，与大自然亲密接触，令身心得到彻底的放松。以疏林草地为特色的酒店及会所前庭构成了私享空间，在这里，你可以呼朋唤友，把酒言欢；动静分区的设计，使之与共享空间完美结合：在享受私密的同时，也能欣赏到外部的自然景色。而酒店及会所后庭则为使用者提供了运动、健身、交流和休憩的独享空间。

Huai'an Guesthouse is located in the north end of Nantai Island in Fuzhou and south to the Nantai Island's Ring Road Park. The meandering Minjiang River was divided into two branches by the island, the south one called Black Dragon River, while the north keeping the original name of Minjiang River. The hotel is situated exactly at the "dragon's head", making the site a truly favorable geographic location.

The geographical advantage has brought the site a unique, beautiful condition. Trees thrive and birds chirrup in the mountains where the hotel is located. In order to reveal the rich natural, historical and cultural of local heritage, designers investigated the site in the mountains carefully and preserved a lot of valuable ancient trees, creating an excellent green setting for the hotel. A large number of historical relics exist in the site, including the ancient county government, the Temple of the Five Ancient Emperors, the Linshui Palace, ancient kiln sites, and

3. 酒店门厅
4. 通过风景优美的小路到达酒店入口

3. Main entrance
4. A path with beautiful views leading to the main entrance

cultural relics of both Zhengchun Weng (one of the best-known scholars in Ming Dynasty) and Xi Zhu (a Confucius philosopher in Song Dynasty), all of which have been well protected and restored in the ensuing constructions. Such rich cultural and natural resources create a cultural atmosphere for the hotel.

The recreational trail, designed to integrate into the hilly terrain of site with parklands, provides a shared space for enjoying the river view and the fragrance of tea, walking in the woods and bathing in the sunlight, that allows visitors to relax and relieve from stress. The parkland and the front garden of the club create private space where the users can gather, drink and chat. With the design of separating dynamic and static elements, the private space which is perfectly integrated with the shared space creates an intimate atmosphere while allowing visitors to have a good view of the external natural surroundings. The backyard garden of the club creates exclusive space where allow people to exercise, work out, communicate, and take a rest.

5. 对称的花钵预示着空间的转换
5. The symmetrical setting of the flowerpots suggests a change of spatial design.

6. 对古树的保护是该项目的主要设计内容之一
6. Preservation of the ancient trees is one of the main components of this project.

① 原始道路　----挖方控制线
② 覆土层　　----文物保护线
③ 文物保护层
④ 回填种植土
⑤ 设计道路

① 建设之前的场地：缺乏规划的野生树种，覆土层与古窑遗址层都离地表很近

② 建设之后的场地：为了保护地下古窑址，在设计需要开挖的区域保证挖方深度在控制线以内，并合理整合植物配置，确保古窑址层的稳定性

7. 以从当地村落收集而来的石材铺筑而成的小路
8. 为了保护地下古窑址文物层，设计开挖区域保证挖方深度在控制线以内
9. 设计保留了古树保护区原始标高，最大限度保护了原生古树

7. The stones collected from local villages are reused for path paving.
8. The depth of construction excavation was limited to protect the ancient kiln relics underground.
9. The project remains the original elevation of the ancient tree area.

项目名称＿福州淮安国宾馆
委托单位＿融侨（福州）置业有限公司
设计内容＿方案及施工图设计
项目面积＿56 000m²
设计时间＿2010年

Name＿Fuzhou Huai'an Guesthouse
Client＿Rongqiao (Fuzhou) Real Estate Co., Ltd.
Service＿Landscape design and construction drawing
Size＿56 000 m²
Year of design＿2010

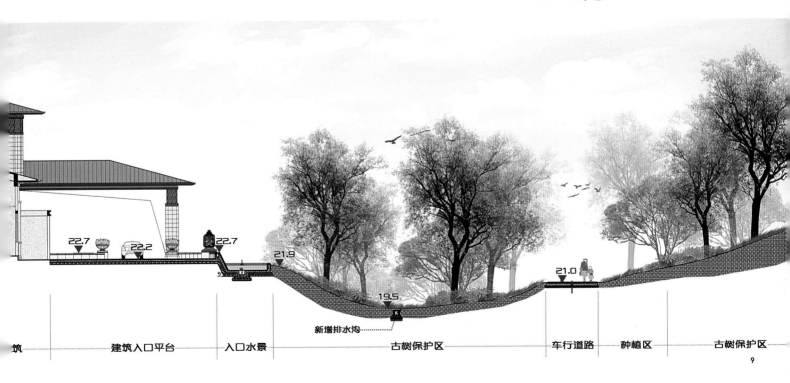

文旅　Cultural Tourism

福州华侨公园
FUZHOU HUAQIAO PARK

华侨公园，顾名思义，就是在福州这座著名的华侨之城，以华侨文化为线索，旨在创造出更多表现华侨文化的场景，以纪念他们为世界各地进步和社会文明所做出的贡献。

一幅世界地图的浮雕成为公园入口的标志性景观，结合草坪上的50根国旗柱，讲述了华侨在全世界的分布区域。在地图延伸的区域中设有一个舵形时钟，象征着海外华侨不断进取拼搏的精神。连接整个公园的元素——场地中的叶形绿岛分布在大草坪上，这一设计象征着海外华侨终归祖国母亲怀抱的落叶归根的文化精神。

整个设计在充分尊重场地人文精神的同时，也重视生态性的营建。树木葱郁的小丛林和散步游径等元素为人们在闹市中提供了一个可以放松身心、亲近自然的悠闲之所。

1. 华侨公园成为隔江远眺的最佳观景点
2. 总平面图

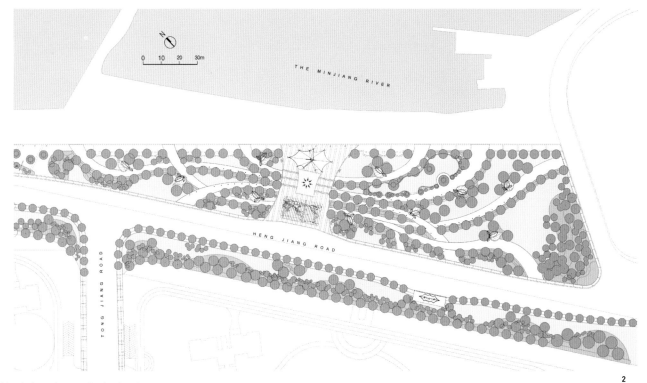

1. The Huaqiao Park provides the best place to enjoy the river view.
2. Site plan

文旅　Cultural Tourism

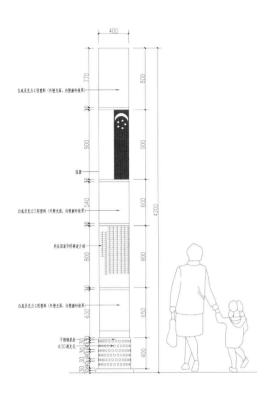
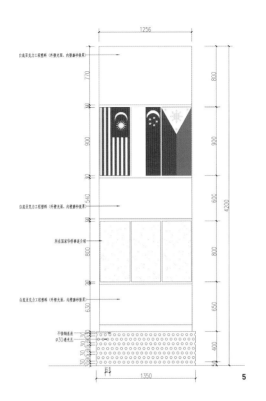

3. 记载着福州华侨概况的国旗柱沿路设置
4. 线性构图在立面上得以延续
5. 国旗柱详图
6. 国旗柱点亮华侨公园,成为闽江江畔的独特风景

3. The national-flag pillars representing the distribution of the global overseas Chinese are set along the park road.
4. The linear composition in the spatial design
5. Detailed plan of the national-flag pillar
6. The lighted national-flag pillars become a unique landscape on the riverfront of Minjiang.

文旅　Cultural Tourism

7. 叶片状膜结构凉亭同样也是福州侨乡"落叶归根"文化精神的象征
8. 膜结构下方以小料石拼成的叶片图案铺地
9. 公园为人们在闹市中提供了一个可以放松身心、亲近自然的悠闲之所
10. 绿树成荫的滨江路

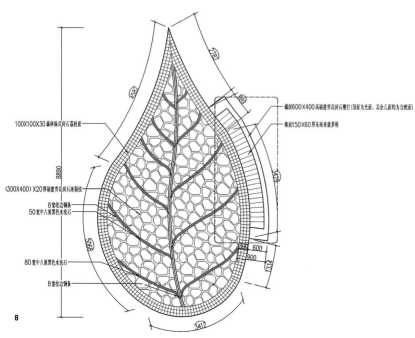

7. The leaf-shaped membrane structure pavilion also implies the culture of the overseas Chinese returning to their mother country.
8. Pavement plan of the leaf-pattern pebble-surface under the membrane structure.
9. The park provides a leisure place in the busy urban area.
10. The shade path along the river

11. 绿荫环绕的儿童游戏空间
12. 重复出现的休闲小空间呈现节奏美感

The Huaqiao Park, or literally Overseas Chinese Park, is located in Fuzhou, a famous city as the hometown for numerous overseas Chinese. Taking the culture of overseas Chinese as a theme, this project aims at creating various scenarios as a window on the culture of overseas Chinese, and commemorating their contributions to the local development and civilization all over the world.

The embossment of a world map acts an iconic landscape at the entrance of the park, representing the distribution of the global overseas Chinese with the 50 national-flag pillars on the grassland. The embossment extends and forms a helm-shaped clock which symbolizes overseas Chinese's spirit of constantly striving for progress. Leaf-shaped green islands as an element connecting the whole park are scattered on the grassland, telling a story of falling leaves settling on the roots and implying the culture of the overseas Chinese returning to their mother country.

The design based on the local cultural characteristics emphasizes on ecological creation. The thriving woods and trails provide a leisure place to relax and connect with the nature in the busy urban area.

11. Trees surround and provide shade for the children playground.
12. A series of repeated small-scale leisure spaces bring a rhythmical experience.

项目名称 _ 福州华侨公园	**Name** _ Fuzhou Huaqiao Park
委托单位 _ 融侨（福州）置业有限公司	**Client** _ Rongqiao (Fuzhou) Real Estate Co., Ltd.
设计内容 _ 方案及施工图设计	**Service** _ Landscape design and construction drawing
项目面积 _ 20 000m²	**Size** _ 20 000 m²
设计时间 _ 2009~2010年	**Year of design** _ 2009~2010

文旅　Cultural Tourism

峨眉山蓝光安娜塔拉度假酒店
ANATARA RESORT IN MOUNT EMEI, SICHUAN

1. 建成后的酒店展现了对设计的高度还原
2. 独具魅力的度假酒店源自泰国安纳塔拉的住享体验

1. The built-up hotel greatly realizes the vision of the design.
2. The design of the glamorous resort was inspired from Anatara Group's Thai style.

文旅　Cultural Tourism

项目位于世界文化与自然遗产、中国四大佛教名山之一的峨眉山山脚下，以峨眉山庄严的寺庙和秀美的自然风光为背景，融入安纳塔拉集团的泰式风情，创造出一种全新的景观特色。对称模纹花坛及台阶水景烘托出庄严的仪式感。各类小品的选择也颇为讲究，与周围环境共同体现出庄重、内敛、低调的景观氛围与浓郁的人文底蕴，充分体现了项目追求休闲、生态、自然的设计宗旨。

The project is located in the foot of the Mount Emei, a famous world-class cultural and natural heritage site which is also one of China's four great Buddhist mountains. The Thai style of Anatara Group was integrated with the surroundings of sacred temples and stunning natural scenery of Emei to create a new landscape characteristic. The symmetrical setting of the carpet flower beds and the terraced water features express a strong ritual sense. All structures were elaborately designed, creating a stately, modest landscape with a rich cultural context that well consisted with the project's design purpose of making a relaxing, ecological, and natural destination.

3. 对称模纹花坛和台阶水景烘托出入口景观的庄
4. 泰式风情的景观营造

3. The symmetrical setting of the carpet flower and the terraced water features express a s stately atmosphere.
4. The Thai style landscape

项目名称	峨眉山蓝光安娜塔拉度假酒店
委托单位	峨眉山蓝光文化旅游置业有限公司
设计内容	施工图设计
项目面积	10 000m²
设计时间	2011年

Name	Anatara Resort in Mount Emei, Sichuan
Client	Emei Languang Cultural Tourism Real Estate Co., Ltd.
Service	Construction drawing
Size	10 000 m²
Year of design	2011

文旅　Cultural Tourism

成都西部文化城

WESTERN CULTURAL NEW TOWN IN CHENGDU

1. 写字楼入口体现出商业办公形象
1. The entrance of the office building creates a business image.

项目位于成都市温江区光华大道南侧,周边道路系统发达,地理条件优越。本项目旨在通过引进不同的文化企业,并尝试通过多种企业文化的碰撞、创新,以汇聚成为成都西部最大的文化产业力量。

在中国传统文化中,庭院总是人们休闲、交流、娱乐的亲切场所。本项目借用院子文化作为设计的景观主题,院内包含"一亩"企业花园——企业文化,欢乐工作之源;"一方"私属净土——梦想家园,闲适生活之泉。院外拥有繁华商业——时尚便捷生活风尚,非正式交流空间。项目结合"院内院外,文化聚场"的核心主题,通过高品质的文化景观,打响"西部文化城"的城市品牌,在提高房地产溢价空间的同时,打造一座名副其实、传承与创新并重的"文化之城"。

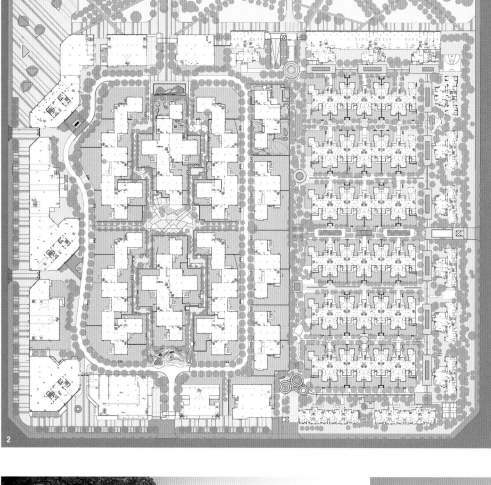

2. 总平面图
3. 墨韵主题景观
4. 花园洋房主入口
5. 栽植樱花的庭院
6. 栽植秋色叶树种的庭院
7. 中轴银杏大道
8. 儿童活动空间
9. 商务办公区中轴广场
10. 夜景

The project is located at the south side of Guanghua Avenue of Wenjiang District, Chengdu, with a well-developed road system connecting with its surroundings. The project aims to become the greatest cultural industry area in west Chengdu by introducing various cultural enterprises to motivate interaction and innovation among different cultures.

In traditional Chinese culture, the courtyard has long been a place of relaxation, communication, and recreation. This project identifies the courtyard culture as the main theme: Inside each yard there is a private club, "Yi Mu," which provides a place for cooperate employees to relax and have pastime, and a private residence, "Yi Fang," which is a refuge from the chaos of modern life. Outside every yard there are busy commercial streets where people can enjoy convenient fashion lifestyle. With a core concept of "cultural cohesion within and beyond the courtyard", the project celebrates an urban brand of "West Chengdu Cultural Area" through creating quality cultural landscapes — a real "Cultural Area" that values heritage and innovation will be built while bringing an increase of the local real estate price.

2. Site plan
3. Ink charm theme landscape
4. Entrance landscape of the garden house area
5. Courtyard with blossoming cherry trees
6. Courtyard with foliage-ornamental trees
7. Ginkgo tree allee
8. Children activity space
9. Central-axis Square of the business-office area
10. Night view

项目名称 _ 成都西部文化城
委托单位 _ 成都市新华创智文化产业投资有限公司
设计内容 _ 方案及施工图设计
项目面积 _ 133 400m²
设计时间 _ 2014年

Name _ Western Cultural New Town in Chengdu
Client _ Chengdu Xinhua Chuangzhi Cultural Industry Investment Co., Ltd.
Service _ Landscape design and construction drawing
Size _ 133 400 m²
Year of design _ 2014

成都璞锦酒店
THE PUJIN HOTEL, CHENGDU

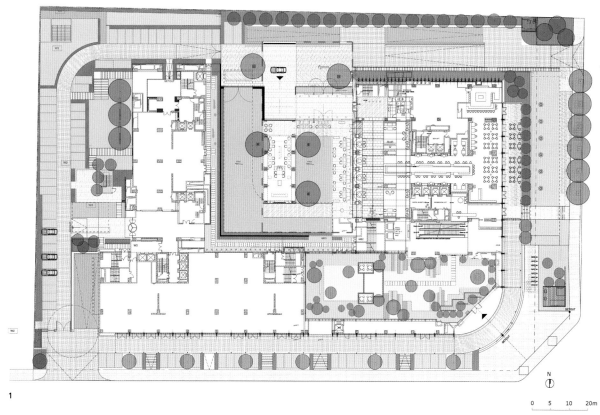

1. 平面图
2. 酒店中庭鸟瞰
3. 绿墙围合的中庭
4. 酒店入口
5. 竹与水——简洁的设计元素

1. Site plan
2. Bird's eye view from the hotel's atrium
3. The atrium with vertical landscapes
4. The entrance
5. Bamboo and reflecting pond — concise landscape elements

璞锦酒店延续国际知名奢华酒店运营商URC酒店管理集团"都会桃源"的概念，将度假胜地酒店的客户体验巧妙地融合到当代都会空间中。

该项目的景观设计在尊重与欣赏场地周边自然环境的同时，充分解读建筑方位、空间及功能，将景观与建筑结合为一体。我们在设计中着重考虑设计空间与自然空间的融合。通过低调的现代设计风格和简洁的设计语言，对日常生活及娱乐的概念进行了全新定义，展现出了现代大都市的独特休闲氛围。与此同时，自然生态的理念一直贯穿设计始终，力图从方方面面体现尊重自然而不仅仅是改造自然的现代设计思想，使人造环境与自然环境密切结合，相互辉映。

The PUJIN Hotel in Chengdu follows the concept of "Arcadia in metropolis" raised by the URC Hotel Management Group, an international famous luxury hotel operator. The project integrates a resort experience into the design of modern metropolitan spaces.

The project respects the natural environments in the surroundings, while integrating landscape design with the architectures of the site with an overall consideration of the location, spatial layout, and functions of the existing buildings. The design highlights the idea of merging the constructed space into the natural context. The project demonstrates a brand new lifestyle and recreation through the modest, modern design with concise elements, creating a distinctive leisure place. Meanwhile, the design employs ecological concept throughout the entire project that is consistent with one of the goals that design is to harmonize with the nature rather than to dominate it.

项目名称 _ 成都璞锦酒店	**Name** _ The PUJIN Hotel, Chengdu
委托单位 _ 英祥集团	**Client** _ Yingxiang Group
合作设计单位 _ 澳洲LAYAN概念方案	**Cooperator** _ LAYAN Design Group Australia (concept design)
设计内容 _ 方案及施工图设计	**Service** _ Landscape design and construction drawing
项目面积 _ 8 654m²	**Size** _ 8 654 m²
设计时间 _ 2014年	**Year of design** _ 2014

福州融侨水乡酒店

FUZHOU RONGQIAO WATERTOWN HOTEL

　　福州融侨水乡酒店依闽江而建。置身酒店，临窗眺望，雄伟的金山大桥、对岸风格各异错落有致的城市建筑、远处连绵的青山相互交融，形成如画的风景。每当夜幕降临，对岸五彩的万家灯火构成的繁荣城市夜景尽收眼底。

　　酒店东面与景色宜人的融侨水乡别墅区紧紧相连，形成一道赏心悦目的风景线。酒店景观设计采用简洁的现代风格，良好地组织了酒店人流和车流交通，并在关键节点上设计了水景和景墙。酒店四周绿树如荫、风景优美，营造出恬静、舒适的世外桃源。

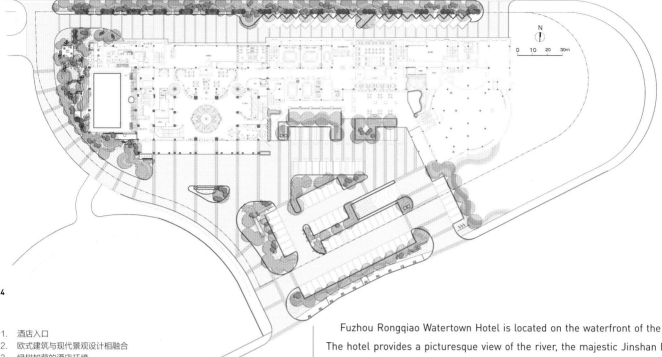

1. 酒店入口
2. 欧式建筑与现代景观设计相融合
3. 绿树如荫的酒店环境
4. 总平面图
5. 酒店外廊
6. 入口水景

1. The entrance
2. The project integrates modern landscape design with the existing European style buildings.
3. Shady trees in the surroundings
4. Site plan
5. Corridor
6. Water features at the main entrance

Fuzhou Rongqiao Watertown Hotel is located on the waterfront of the Minjiang River. The hotel provides a picturesque view of the river, the majestic Jinshan Bridge, and the glamorous city skyline with a background of rolling hills in distance. When night falls, the hotel becomes the best place to enjoy the colorful lighting and prosperous nightscape from the other river bank.

The hotel connects with the Rongqiao Watertown Villa area at the east, forming an aesthetically consistent landscape. The landscape design for the hotel adopted concise, modern style and well organized the pedestrian and vehicle circulations. Water features and landscaped walls were designed at the key nodes. A beautiful, elegant, and restful environment is created with shady trees in the surroundings.

项目名称 _ 福州融侨水乡酒店
委托单位 _ 融侨（福州）置业有限公司
合作单位 _ ACLA
设计内容 _ 施工图设计
项目面积 _ 6 654m²
设计时间 _ 2011年

Name _ Fuzhou Rongqiao Watertown Hotel
Client _ Rongqiao (Fuzhou) Real Estate Co. Ltd.
Cooperator _ ACLA
Service _ Construction drawing
Size _ 6 654 m²
Year of design _ 2011

成都青城郡五行仙谷度假温泉

FIVE ELEMENTS FAIRYLAND VALLEY RESORT & SPA IN QINGCHENG COUNTY, CHENGDU

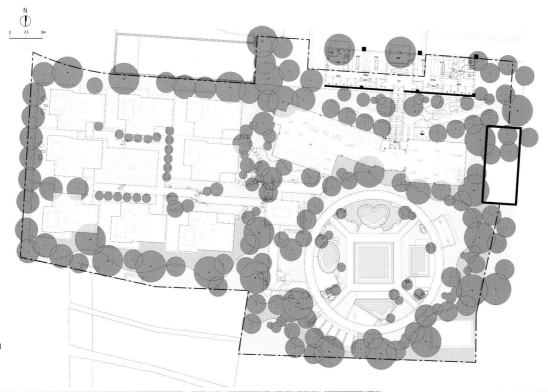

1. 总平面图
2. 温泉入口特色景观墙
3. 云纹图案铺装，烘托入口氛围

1. Site plan
2. Landscape wall at the entrance area
3. Cloud-pattern paving

青城郡五行仙谷度假温泉项目沿袭青城山千年道医文化，集泡汤、膳食、理疗和养生等项目于一体，将道教文化中的祥云文化、洞天福地、龟鹤文化、易经五行、八卦九宫文化融入景观中，打造出青城山下以道教文化为背景的温泉养生圣地。

（1）道意祥云

祥云是道教文化中的一种典型元素。景观设计师在入口处以水景结合弧形景墙进行立面造景，并在入口铺装上设计了祥云纹样，使游客可以脚踏"祥云"进入到温泉服务中心，为度假温泉增添了神秘色彩。

（2）洞天福地

洞天，道教用以称仙人所居之处。青城山是中国道教的发源地之一，被道教列为"第五洞天"。因此，该项目在景观设计上以假山石为造景元素，在汤池入口处叠石造洞营造出别有洞天的景观氛围。

（3）龟鹤齐龄

在中国传统文化中，龟、鹤均有长寿之意。在SPA庭院中的水景中央设计龟鹤雕塑，寓意长寿延年。

（4）五行养生池

五行养生池结合"金木水火土"五行文化打造自然养生会所，根据五行方位与属性定制特效养生汤池，分别对应不同的养生功效。

太极图是道家宇宙观和方法论的模式图，象征着生命的生生不息。设计师在凉亭的设计中打造了太极阴阳鱼中心，并围绕五行养生池设计了融合八卦元素的园路，为项目增添了浓厚的文化氛围。

The project follows the culture of Taoist medicine that has developed in the Qingcheng Mountain area and has a history of thousands of years. Programs such as hot springs, restaurants, SPA and physical therapies are integrated into this project. By abstracting from the cultures of auspicious clouds, cave paradise, longevity of turtle and crane, I Ching and Five Elements, as well as the Eight-diagrams and the Nine Palaces, the landscape design aimed to create a healthy Taoism destination of resort and SPA at the foot of the Qingcheng Mountain.

(1) Auspicious Cloud

Auspicious cloud is a typical element in Taoism culture. Designers created a vertical landscape by combining waterscape with arc landscape wall at the entrance area. The cloud-pattern paving was designed to create a mysterious ambience that allows visitors mount "clouds" when they walk into the building.

(2) Cave Paradise

Cave paradise stands for the residence of immortal in Taoism. Qingcheng Mountain is one of the origins of Taoism in China and known as the Fifth Cave Paradise in China. Rockeries were used in the landscape design for the entrance area of the hot springs to create a fairyland hidden by "mountains" with heavy mist.

(3) Longevity of Turtle and Crane

Both turtle and crane have an implied meaning of longevity in traditional Chinese culture, particularly in Taoism. The images of these two creatures are shown as sculptures at the SPA courtyard.

(4) Five Elements Hot Springs

The culture of Five Elements — gold, wood, water, fire and soil — was embodied in the landscape design of the hot springs. Different pools with special effects were customized according to the positions and attributes of the Five Elements.

Tai-chi is the ideograph that reflects the Taoism cosmology and methodology, and implies the endless cycle of the life. A Tai-chi pattern pavilion and winding paths with symbols of Eight-diagrams were designed to respond and accent the Taoism culture in the overall project.

4. 休闲庭院和情景化小品
5. 温泉药酒廊架
6. 五行养生池特色凉亭

4. Courtyard
5. Medicinal liquor corridor
6. Pavilion over the hot spring

项目名称 _ 成都青城郡五行仙谷度假温泉
委托单位 _ 四川炎华置信实业（集团）有限公司
设计内容 _ 方案及扩初设计
项目面积 _ 3 000m²
设计时间 _ 2014年

Name _ Five-Elements Fairyland Valley Resort & Spa in Qingcheng County, Chengdu
Client _ Sichuan Yanva Real Estate Development (Group) Co., Ltd.
Service _ Landscape design and preliminary drawing
Size _ 3 000 m²
Year of design _ 2014

陕西省西咸新区丝路原点小镇

STARTING-POINT TOWN OF THE SILK ROAD IN XI-XIAN NEW DISTRICT, SHAANXI

1
N

1. 总平面图
2. 生态产业经济发展框架
3. 丝路沙洲月牙泉景观再现

1. Site plan
2. Eco-agriculture development framework
3. Rendering of the restored New-moon Spring scenery

在"一带一路"战略复兴伟大中国梦的宏大背景之下，丝路原点小镇选址于西咸新区泾河新城组团，东临泾阳县城，西侧是建设中的泾河新城，规划总面积178hm^2。

小镇总体规划为"一带五区"：一带指丝路风情商业带，这条南北走向的轴线是小镇的核心景观区，其将南侧的形象入口、七星百姓广场、丝路商业街等空间串联起来，结合唐代驿站景观并融合趣味历史故事，再现了唐代丝绸之路的动人画卷。这条文化轴线的北段是海上丝路公园，通过营建环境优美的生态公园，打造海上丝路文化景观，使游客在山水田园间，领略海上丝绸之路的文化魅力。

■ 生态产业经济发展模式

构建生态小镇：
社会·产业·空间·人

> 生态是生物在一定的自然环境下稳定生存和可持续发展的状态。在空间布局由一带五园构成的丝路田园小镇里，空间的分布是产业划分的布局，空间的组合又是社会构成的因素。在社会与产业构成的生态圈里，人是空间构成的核心，人是社会组成的要素，也是推动产业升级的动因，产业的升级促进了社会的进步，社会的发展演绎了人类的文明。

"五区"是指小镇规划的五个功能板块区，分别为丝路田园风情园、青少年农科园、老龄化社会事业园、社会事业服务园和科技农业示范园：①丝路田园风情园依托现有生态农业与传统农耕文化，结合园内的古宅民居，并通过结合丰富的物质遗存，引入农业观光旅游理念，形成独具地域文化特色的农业体验。②青少年农科园以多元的文化体验、丰富的运动休闲设施和场地，打造高水准的青少年训练营地。与此同时，通过建立现代农业科技成果展示平台，向青少年传播农业科普知识和种植技术。③老龄化社会事业园引入优质医疗资源，结合田园自然属性，打造硬件一流、环境舒适的养老生活社区。④社会事业服务园建设新型居住社区，将自然田园风光融入现代生活服务，并同时开发时尚现代的社区综合商业，开展丰富多彩的社区活动，实现田园宜居梦想。⑤科技农业示范园以高效农产品孵化为特色，以农产品采摘为亮点，以市场需求为向导，以休闲观光、体验为载体，是集现代农业种植生产示范、加工、销售、观光旅游为一体的综合性科技农业示范园。

文旅 Cultural Tourism

4. 田园农业观光园
5. 非物质文化遗产博物馆
6. 由传统民居改造而来的精品酒店

4. Eco-agricultural Sightseeing Park
5. Intangible Cultural Heritage Museum
6. Resort reconstructed from old traditional courtyard houses

Under the background of "One Belt and One Road" initiatives advocated by the Chinese central government, the Jinghe new development area was designated as the site of the Starting-point Town of the Silk Road, covering an area of approximately 178 hm².

In the overall plan a framework of "One Belt and Five Zones" is identified. "One Belt" refers to the Silk Road Business and Commercial Belt which forms a north-south axis of the core landscape area, connecting with the south entrance and Seven-Star Citizens' Square. The courier station scenery in Tang Dynasty and best-known legends are integrated into the landscape design, generating a vibrant vision of the ancient trail. A Silk Road waterway memorial park will be built in the northern section of this axis to tell the historic and cultural stories in a modern, ecological way.

"Five Zones" refers to the five theme blocks, including the Eco-agriculture Sightseeing Park, Agricultural Science Park for teenagers, Recreational and Living Area for the seniors, Social Service Area, and High-tech Agricultural Demonstration Area. I Eco-agriculture Sightseeing Park: by taking advantages of the existing ecological and traditional agriculture farming resources and combing agricultural sightseeing with the ancient town's rich heritages, the park provides visitors a unique urban farming experience. II Agricultural Science Park for teenagers: as a high-level teenager training base, the park consists of diverse experience places, and sports and leisure facilities, while being a popular educational center of modern agriculture and planting technologies. III Recreational and Living Area for the seniors: located in beautiful natural setting, the park introduces medical service and recreational facilities and amenities, to create a comfortable living neighborhood for the elderly. IV Social Service Area: this area is supposed to house several new residential developments and a business and commercial complex, as well as various places for community activities. V High-tech Agricultural Demonstration Area: this area features market-oriented agricultural production, combining with recreation, leisure, and tourism services, to build an integrated agricultural demonstration area for the new town.

项目名称 _ 陕西省西咸新区丝路原点小镇
委托单位 _ 泾华投资有限责任公司
设计内容 _ 规划与景观设计
项目面积 _ 178hm²
设计时间 _ 2015年

Name _ Starting-point Town of the Silk Road in Xi-Xian New District, Shaanxi
Client _ Jinghua Investment Co., Ltd.
Service _ Planning and Landscape design
Size _ 178 hm²
Year of design _ 2015

文旅　Cultural Tourism

达利·天屿湖休闲国际社区
DALY TIANYU LAKE INTERNATIONAL COMMUNITY

天屿湖位于湖北省武汉市，湖面300余hm²。这一大型文化旅游地产项目集文化、饮食、娱乐、养生、商业等功能于一体，核心板块为高端住宅、度假酒店、高尔夫会所、体育会所、水上世界及主题乐园。

规划将项目湖西区域打造成集商业、旅游、居住为一体的旅游休闲度假区，其中高尔夫会所设计充分结合原有自然景观和地貌，利用高大乔木与高尔夫果岭营造低调而开敞的高品质景观环境。体育会所则以道路和植物来划分户外空间。羽毛球场及室内泳池等体育健身设施的建设倡导了健康、轻松的生活模式。

湖东区域旨在打造成集休闲、活动、运动为一体的休闲生活公园。湖水作为整个项目的核心资源，景观设计从水中获取灵感，对水纹进行抽象和变形，形成自然、流畅的平面构图，用湖水柔和与缓慢的流动来体现场地闲适的"慢"氛围。

项目交通系统完善，主要由外环车行道、观光车游览道、自行车骑游道和游步道、漫步小径组成。其中东外环道路总长约4.5km，为项目的主要车行景观道，设计以"十二花神"为主题，打造了一条具有文化底蕴的景观大道，植物空间舒适自然，以变化起伏的地形及多层次的植物群落呈现丰富的景观道路空间，为住户提供多样化的住享体验。

1. 海棠春坞景区
2. 天屿湖总平面图
3. 琴音抚柳景区

1. The dock featuring the planting of *Malus spectabilis*
2. Site plan
3. The dock featuring the planting of willows

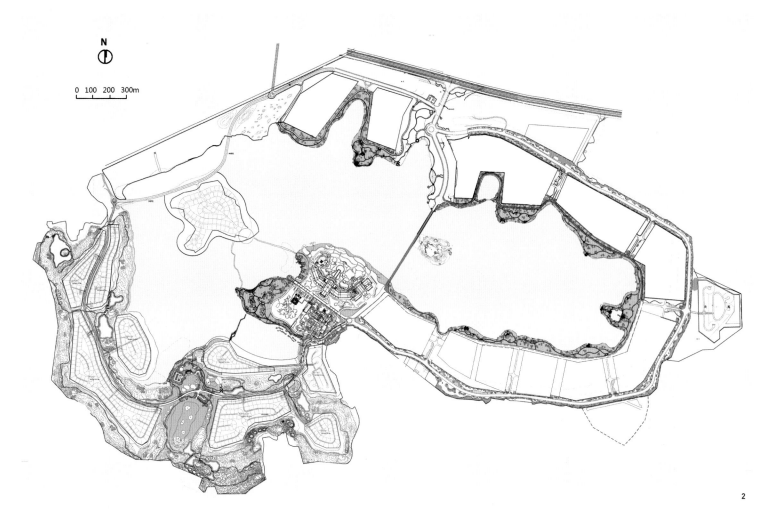

文旅　Cultural Tourism

Located in Wuhan City, Hubei Province, Tianyu Lake covers a total water area of over 300 hectares. This development project involves various programs ranging from cultural tourism to real estates, providing recreational, environmental, and commercial services. It consists of high-end residential areas, resort hotels, a golf course, sports clubs, a water playground and other theme parks.

The western side of the Tianyu Lake is designed as a resort with commercial, touristic and residential properties. The design of the golf course adapts the existing topography while manifesting the natural landscape. Open space for sports clubs is divided by roads and plants. The badminton courts, an indoor swimming pool and other fitness facilities help promote a healthy, relaxing lifestyle.

The eastern side of the lake is designed to create a park for recreational activities and sports programs. Inspired from the core element, the water body, the design team develops a natural layout by abstracting the image of water wave to reflect the quiet, calm atmosphere of the site.

The traffic system mainly consists of the outer ring road, sightseeing bus routes, bike lanes, hiking trails, and tracks. With the theme of "Twelve Floras", the eastern outer ring road acts as the main landscape axis for vehicles with a rough length of 4.5 km. The undulating terrain and multi-layered plant communities creates rich landscapes along the road, offering changing experience for residents.

4. 休闲公园
5. 玉玙阁
6. 天屿湖别墅

4. Park with a traditional Chinese style pavilion
5. Yuyu Pavilion
6. Tianyu Lake Villa

项目名称 _ 达利·天屿湖休闲国际社区
委托单位 _ 湖北达利地产有限公司
设计内容 _ 方案及施工图设计
项目面积 _ 538 295m²
设计时间 _ 2015年

Name _ Daly Tianyu Lake International Community
Client _ Hubei Dali Real Estate Co., Ltd.
Service _ Landscape design and construction drawing
Size _ 538 295 m²
Year of design _ 2015

公共　　PUBLIC SPACE

重庆融侨公园
CHONGQING RONGQIAO PARK

1. 重庆主城内主要的开放式公园
1. The park is one of the main public parks in the city center.

融侨公园项目位于重庆长江南岸的铜元局片区，占地面积约13hm²，北临长江，背依南山，紧邻城市观光休闲景观大道——南滨路。这一片区曾是创办了重庆第一家本土工业企业、引进了重庆市第一台国外设备、点亮了重庆第一盏电灯的重庆铜元局（新中国成立后为重庆长江电工工业集团有限公司）的所在地。1905年4月14日，重庆铜元局设立。铜元局作为重庆开埠后第一个近代工业企业，标志着重庆由农耕文明向工业文明的转折。铜元局后在军阀混战时期被改建为军工厂，并于1937年被正式更名为"兵工署第二十工厂"。由于铜元局地处江边，双峰山、鹅岭夹峙长江，地势险峻，日军飞机多次轰炸未果。工人们夜以继日地为前线输送枪弹，为抗战胜利做出了巨大贡献。

设计团队以当地深厚的历史文化和卓越的产业发展为切入点，希望在保持场地的历史源流与时代特色的同时，满足周边居民日益增长的休闲需求。设计以"生态、文化、休闲、家园"为主题，以"一湖"（湿地）为核心，以"两廊"（文化休闲走廊和生态休闲走廊）为框架，构建了五大功能分区：运动休闲区、铜元局历史文化区、亲水休闲区、迎宾广场区和生态湿地休闲区。

景观森林设计实践

在融侨公园的景观设计中，水的应用是核心，也是点睛之笔。我们因地制宜地利用湿地的优势，建造了一处面积达4hm²的内湖，并巧妙地引水入林，令处处皆有水景。

设计力图在打造一个可供市民休闲、娱乐的公共环境的同时，亦通过对往昔的追溯和雕刻来展现场地历史。随着时代的变迁，铜元局遗址中一些记录了历史印记的老建筑和设备早已被拆除。融侨公园用现代设计语言重塑了场所记忆和精神，通过符号化的处理，串联起整个场地的历史信息。例如，反映铜元局铸造历史的"货币广场"、反映长江电工工业时代的"雕塑广场"等，使历史记忆得以再现。

作为"森林重庆"规划的十大公园之一和重庆主城区内最大的免费公园，建成后的融侨公园提升了整个南岸区生态环境的宜居性，结束了几十年来铜元局片区无公园的历史，为当地百姓创造了舒适的休闲与游憩场所，并成为重庆市的标志性公共景观，受到了市民和政府的多次赞扬。2010年4月23日，重庆融侨公园项目在第三届"天府杯"优秀园林景观设计（工程）大赛中获得金奖，并在2010"重庆最美公园"评选活动中荣获"十大最美公园"称号。

2. 总平面图
3，4. 长江电工工业为该地块留下了一段火热的工业记忆
5. 草坪栈道与小径

2. Site plan
3,4. Chongqing Yangtze Electric Industry left a rich industrial heritage on the site.
5. Boardwalk and path

6. 黄昏下的融侨公园
7. 穿梭于花木间的木栈道
8. 水杉区域剖面图

6. The park in dusk
7. Boardwalk winding in blossoming shrubs
8. Sections of the dawn redwood grove (*Metasequoia* spp.)

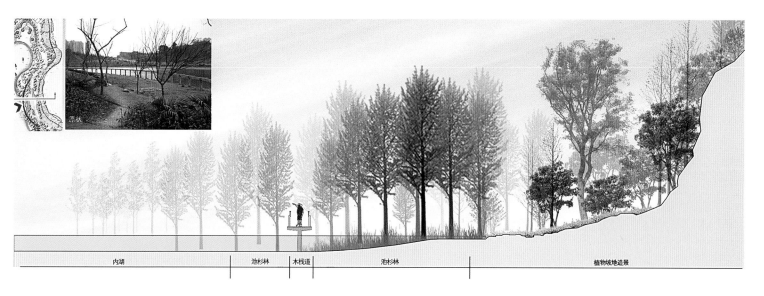

内湖 | 池杉林 | 木栈道 | 池杉林 | 植物坡地造景

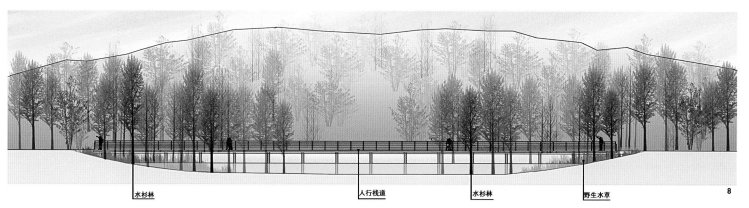

水杉林 | 人行栈道 | 水杉林 | 野生水草

公共　Public Space

Covering an area of approximate 13 hm^2, Rongqiao Park is located in the Tongyuanju District in Chongqing, at the south of the Yangtze River, with the Nanshan Mountain at the back. The site is also adjacent to the Nanbin Road which is a landscape boulevard along the south riverfront. The industrial campus in which the park located was the former site of the Chongqing Mint — known as the Chongqing Yangtze Electric Group after the founding of new China. Founded on April 14, 1905, the Mint was the first local industrial plant in Chongqing and the first factory that introduced modern equipments from aboard, which was a turning point in Chongqing's history from the agricultural civilization towards the industrial era. The Mint was employed as an ordnance factory in the warlord period and formally renamed as "The Twentieth Ordnance Factory" in 1937. Because of its topographically defensible location, the factory was one of the last survived places in Japanese air raid and produced bullets to

supply the frontline, significantly contributing to the victory of the war.

Taking a comprehensive study of the rich historical heritage and local industrial development as the starting point, a plan was proposed that highlights the site's glorious past and meets the citizens' increasing needs for recreation. The design identified "One Lake" (the wetland) as the core and "Two Corridors" (the culture corridor and leisure corridor) as the backbone, to form five main areas: sports and leisure area, Mint history area, waterfront area, entrance area, and the ecological wetland area.

The water body is one of the key elements in the landscape design of the park. Adapting to the natural terrain of the wetland, the design formed a 4 hm² lake and conveyed the water to the forest to create various water features.

How to highlight the cultural history of the site was another challenge to the designers. Since many historical buildings and structures had been demolished over the past century, the design used modern landscape language to revitalize the memory

9. 跌水空间
10. 湖边草坪成为周末家庭聚会的最佳空间
11. 植物、置石、小径通过巧妙结合，创造出怡人的散步环境

9. Flowing cascade
10. The lawn on the lakeside becomes a wonderful place for family gathering.
11. Pleasant pedestrian environment is created through carefully choreographed plants, stones, and paths.

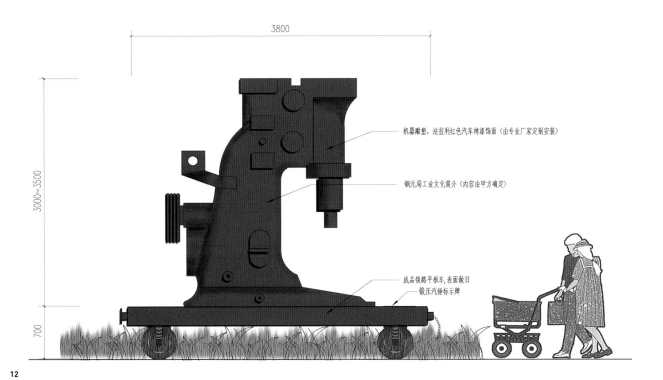

12

13

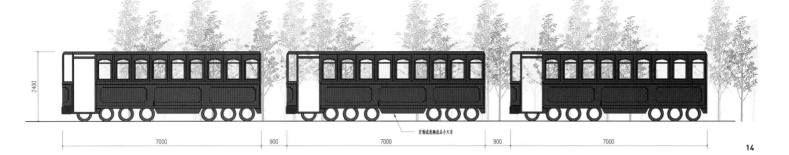

14

and the place spirit by contacting the whole site with symbolic elements that tell stories about the impressive past, such as Currency Square and Sculpture Square that respectively recall the years of the Chongqing Mint and the Yangtze Electric Industry.

As a part of the Forest Chongqing plan and the largest free, public park in the city center, Rongqiao Park improves the livable and ecological environments of the local district and ends the park- absent history of the industrial area. It provides a popular, pleasant recreational place for local people, while becoming a landscape mark in both district and city scale. The project won a gold medal in the Third "Tianfu Cup" Landscape Design (Engineering) Competition on April 23, 2010 and was awarded the Top Ten Most Beautiful Park in Chongqing in 2010.

15

12~14. 场地上的工业印记
15. 虚实结合的工厂景象带领游客走入往昔记忆

12~14. Industrial memories of the site
15. An epitome of the industrial scenario recalling memories of the glorious past

公共 Public Space 71

16. 山城、雾都，缩影在14hm²的融侨公园中
17. 通过植物生态净化系统将长江江水引入内湖，形成了一片美丽的水上植物区
16. The natural and cultural scenery of Chongqing, and the mountainous and foggy city, is well represented in the park.
17. An enchanting aquatic plant area is formed through conveying water of the Yangtze River into the lake after filtered by a wetland eco-purification system.

项目名称 _ 重庆融侨公园	**Name** _ Chongqing Rongqiao Park
委托单位 _ 融侨长江(重庆)房地产有限公司	**Client** _ Rongqiao Yangtze River (Chongqing) Real Estate Co., Ltd.
设计内容 _ 方案及施工图设计	**Service** _ Landscape design and construction drawing
项目面积 _ 133 400m²	**Size** _ 133 400 m²
设计时间 _ 2009年	**Year of design** _ 2009

福州淮安环岛路公园
FUZHOU HUAI'AN RING ROAD PARK

1. 看似冲突的现代渡口与古渡头形成了一种当代与历史的"融合"
1. The visual confliction between the modern quay and the ancient ferry creates a bond with the rich history of the site.

淮安位于福州南台岛北端，闽江在此被南台岛一分为二，南港称乌龙江，北港仍称闽江（亦称"白龙江"）。

福州淮安环岛路公园项目占地面积约91 000m²，东、西、北三面环水，南面为规划中的三环路，且紧邻福州大学城。场地文化底蕴浓厚，古渡口、接官道、相公庙、大桥头等都是旧时遗留下的宝贵文化遗产。场地中的原始地貌和原生植被保存情况良好——场地中现存完好的493棵原生龙眼古树更是让人惊叹。当设计师们第一次踏上这块滨江绿地时，便被这里的景色深深吸引。结合淮安的历史文化和场地的特殊性，设计师在交通流线的设计上最大限度地保留了场地中的原始植被和滩地，旨在为该地区营造一系列文化韵味浓厚、闲适宁静的景观空间，重现芋原古渡、"三江"胜迹等景致。

（1）芋原古渡

芋原渡口从宋代起就是中国与日本进行陶瓷贸易的一个重要码头。渡头一株参天古榕，屹立在江边已四百余年。古时，福州秀才进京赶考时，须经过古渡口上船，而中举返乡之时亦须路过这里。由于登船时的低头动作犹似参拜祈福，祈祷家族的世代荣耀与平安，直至今时，当地居民仍保有在这株古榕树下祈福的习俗。相传大儒朱熹也曾在芋原古渡附近的驿站留墨"芋原"二字。

（2）"三江"胜迹

闽江至福州淮安后被辟为二支，白龙江、乌龙江两江绕流福州南台岛，而后汇合成马江东流入海。这就是福州的"三江"胜迹。白龙与乌龙两江如两条蛟龙一般守护着这片褪尽尘嚣的古邑。当地传说成为了设计师的灵感源泉，"二龙戏珠"的浮雕也被形象化地镶嵌在场地之中。

Huai'an, the north end of Nantai Island in Fuzhou, divides the Minjiang River into two branches: the south one called Black Dragon River, while the north keeping the original name of Minjiang (also known as White Dragon River).

The Huai'an Ring Road Park covers an area of around 91 000m^2, surrounded by water at three sides with a proposed ring road and a university campus at the south side. The site has a glorious cultural history with a large number of heritages including the ancient ferries, Jieguan Road (a road for welcoming governmental officials), Xianggong Temple (a Qing Dynasty structure), and the Grand Bridgehead. The original natural environment and vegetation conditions are well preserved on the site, especially 493 ancient longan trees have been perfectly remained here through hundreds of years. The designers were immediately attracted by the local scenery when they first arrived on the island. In terms of the design of pedestrian and vehicle circulations, designers decided to maximize the preservation scope of the native vegetation and original foreshores while creating attractive, comfortable, and peaceful landscape spaces and restoring historical features such as the Ancient Yuyuan Ferry and the View of Three Rivers.

2. 从栈桥望去,可一览渡头景色
3. 项目重现了芋原古渡的历史风貌

2. The view from the trestle bridge
3. The project restores the historical feature of the Ancient Yuyuan Ferry.

公共　Public Space

(1) Ancient Yuyuan Ferry

Yuyuan Ferry had been an important dock for China-Japan pottery and porcelain trade since the Song Dynasty. A large banyan tree has been standing at the Ferry for over 400 years. In the ancient time, scholars leaving Fuzhou for the capital to attend the imperial examination would board the boat from here; they would land here again when they passed the examination and returned home. The postures and body movements when the scholars were boarding were like praying and showing their respect to deities for glory and safety for their family. Until now the local people still come to this ancient banyan tree and pray for blessing. It is said that Zhu Xi, a famous ancient Chinese philosopher, once wrote down the two Chinese characters of "yu yuan" in an ancient roadhouse near the Ferry.

(2) View of Three Rivers

Minjiang River is divided into two branches — Black Dragon River and White Dragon River — by the Nantai Island and the two flow around the island and converge into the Majiang River that runs eastward into the sea. A spectacular, beautiful view thus is formed by the three rivers. Furthermore, the legend of the two branches like two dragons protecting the ancient town, inspired the designers, and an embossment of "two dragons playing with a pearl" is created on the site.

项目名称 _	福州淮安环岛路公园
委托单位 _	融侨（福州）置业有限公司
设计内容 _	方案及施工图设计
项目面积 _	91 000 m²
设计时间 _	2009~2010年

Name _	Fuzhou Huai'an Ring Road Park
Client _	Rongqiao (Fuzhou) Real Estate Co., Ltd.
Service _	Landscape design and construction drawing
Size _	91 000 m²
Year of design _	2009~2010

4,5. 雨水冲刷后的园路景观：园路设计大胆加入"红"元素，与周围植物的"绿"搭配出令人耳目一新的效果

4,5. The view along the path after rainfall. The design of the paths in the park boldly uses the red pavement as an attractive element, creating a refreshing landscape with a delicate design of vegetation.

公共　Public Space

福州南江滨公园
FUZHOU NANJIANGBIN PARK

1. 总平面图
1. Site plan

福州是一个人文荟萃的有"福"之地。闽江流经福州的50多公里，两岸有着许多人文古迹和自然景观；福州又是中国著名的侨乡，祖籍福州的华侨、华人多达300余万人，他们从这里出发，又回望故里。而在福州突飞猛进的城市发展中，闽江南岸则相对发展较慢。如何对闽江南岸进行改造，建设更美的家园，让离开这里的人们能寻找到家的记忆；又如何将闽江厚重而悠久的文化在新的时代中有新的呈现，是我们在这个项目中需要探索的话题。

根据政府规划，未来的南江滨会被建成集商业金融、娱乐休闲、观光游览、商务办公、文化教育和居住等多种功能于一体的都市活力长廊，集中展示福州的历史风貌、时代韵律与开放精神。所以我们在景观设计中，以"活力南江滨"为主题，打造新时代南江滨的生活场景，设置融侨广场、福文化广场、老福州风情街等一系列主题性景观节点，共同构成以福州地域文化为主要脉络的景观序列，充分展现南江滨充满活力的精神气质。

（1）融侨广场

著名爱国华侨林文镜先生1989年于福州创办融侨集团，始终秉持"为居者着想，为后代留鉴"的经营理念，以"构建理想城市生活"为己任。融侨广场中心最醒目的一艘船形雕塑上镌刻的"1989"，意指融侨集团于1989年在福州正式成立。沿着滨江步道，人们可以看到依主题顺序排列的景观灯柱，高低错落的绿岛和块状绿地更加充实了此处的空间。

（2）福文化广场

"一口田，衣禄全"，"福"在中国传统文化中有着特殊的含义，它蕴含着中国人追求幸福的美好心愿；而福州因"北有福山"而得名，又有"福海宝地"之盛赞。以"福"为主题的福文化广场，既以符号的形式表达了人们对幸福的渴求，也寓意融侨集团和福州同福同源的深厚情谊。

（3）老福州风情街

老福州风情街展示了闽都的各类文化项目。南江滨滨水而建的文化长廊成了福州向众人展示自身特色的一个重要名片。

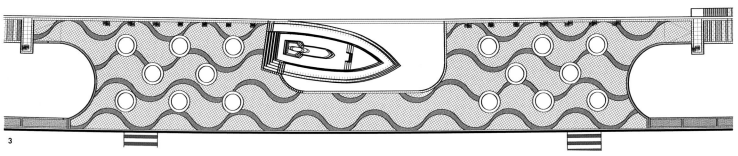

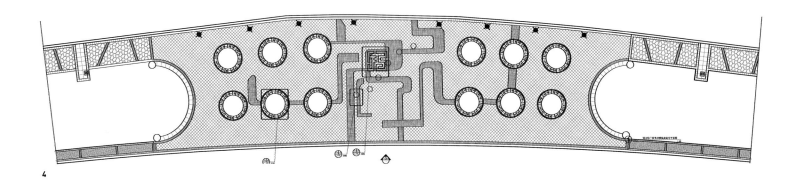

2. "1989"船形雕塑昭示着融侨集团于1989年在福州成立的历史
3. 融侨广场总平面图
4. 福文化广场平面图
5. 怡人的林下空间
6. 福文化广场上的立方体福字雕塑

2. A "1989" was engraved on the hull of a boat sculpture in the park, commemorating the year when the Rongqiao Group was founded in Fuzhou.
3. Site plan of the Rongqiao Square
4. Site plan of the Blessing Culture Square
5. Shade space
6. "Blessing" sculpture in the Blessing Culture Square

公共 Public Space 83

我们还将高架桥下的空间尽可能地利用起来：一部分展现老福州的民俗文化，如闽剧广场，本地人在这里嬉戏玩乐，外地游客在这里拍照留念；一部分则作为功能性场地——停车场使用，为驾车到南滨江公园游玩的人们提供更为便利的服务。

（4）滨江休闲生活广场

无论是晨光初照的清晨，还是霓虹闪烁的傍晚，在南江滨，都能看到嬉戏的孩子、闲适的老人、热恋的男女，他们在这里释放运动的激情，或与朋友、家人分享生活的愉悦……一种崭新的生活形态——"公园生活"已经潜移默化地成为闽江南岸人们的日常生活形态。

（5）华侨公园

华侨公园，顾名思义，就是在福州——这座著名的华侨之城，以华侨文化为线索，创造出更多可以了解华侨文化的场景，以纪念他们为世界各地进步和社会文明所做出的贡献。一幅世界地图的浮雕成为了公园入口的标志性景观，结合草坪上的50根国旗柱，讲述了华侨在全世界的分布区域。在地图延伸的区域中设有一个舵形时钟，象征着海外华侨不断进取拼搏的精神。连接整个公园的元素——场地中的叶形绿岛分布在大草坪上，这一设计象征着海外华侨终归祖国母亲怀抱的落叶归根的文化精神。

整个设计在充分尊重场地的人文精神的同时，也重视生态性的营建。树木葱郁的小丛林和散步游径等元素为人们在闹市中提供了一个可以放松身心、亲近自然的悠闲之所。

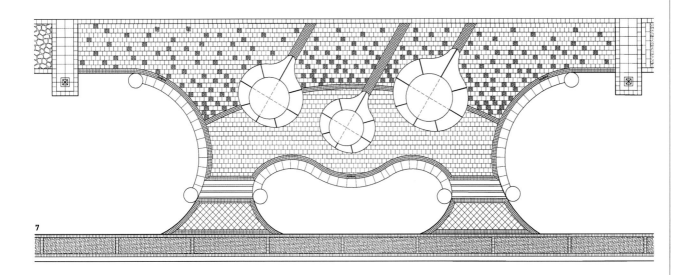

7

Fuzhou is a blessed land with teeming cultures. Numerous historic heritages and natural landscapes are distributed along the 50-kilometer-long banks of the Minjiang River which flows throughout the Fuzhou City. Fuzhou is also characterized as the most famous and largest hometown for overseas Chinese, being the ancestral home for over three million Chinese residing abroad. The urban development of the south bank of Minjiang River is considered quite slow in contrast with the rapid urbanization happened on the north side. How to build the south bank of the Minjiang River towards a more beautiful home and help people who left find their memories of hometown? How to demonstrate and translate the rich cultural heritage of Minjiang River in this new age?

In the urban planning vision proposed by the local government, the South Riverside will be built as a city gateway and a vibrant corridor that integrates a variety of functions including commercial space, business offices, recreational places, touristic and cultural attractions, as well as educational and residential areas, showing the historic features and contemporary spirit. So we identified the design theme as "The Vibrant South Riverside" to create livable places that consist of a series of landscape nodes, such as Rongqiao Square, Blessing Cultural Square, and Traditional Culture Street, to form landscape sequence that features the local geographical and cultural characteristics.

(1) Rongqiao Square

Well-known patriotic overseas Chinese Mr. Lin Wenjing founded Rongqiao Group in 1989 in Fuzhou. The company has always been stick to its philosophy of "caring for the contemporaries and exemplifying for future generations", and deems "building ideal urban life" as its persistent duty. A "1989" was engraved on the hull of a boat sculpture in the center of the square, commemorating the year when the Rongqiao Group was founded. The riverfront trail provides people a great place to view the landscape lampposts in distance. Scattered vegetation

7. 以司南为主题的节点平面图
8,9. 司南形状的树池

7. Site plan of the Sinan (compass in ancient China) Theme Square
8,9. Sinan-shaped planting beds

公共　Public Space　85

clusters and massive green space along the trail enrich visitors' experience.

(2) Blessing Cultural Square

As the saying goes "with farmland the family has everything it needs for living" (the literal meaning of the writing Chinese character of Fu, meaning "blessing"). "Blessing" has a wonderful implication in traditional Chinese culture. It contains the Chinese people's good wishes to pursue happiness; Fuzhou was named after the Fu Mountain in the city north and has been praised as the "blessing and treasure place." Blessing Cultural Square expresses people's desire for happiness in the form of symbols and implies the metaphorical meaning of interdependent fate between Rongqiao Group and Fuzhou City.

(3) Traditional Culture Street

The Traditional Culture Street acts as a window to demonstrate the rich traditional cultural features of the Fuzhou City. A cultural corridor along the riverfront creates an important image for the city by showing various local characteristics. We also utilized the space under overpasses to transfer them into folk places (such as Fujian Opera Amphitheater) and car-parking area.

(4) Riverside Leisure Plaza

Whether it is a sunny morning or a neon-flashing night, the South Riverside Park is always a popular place and you can see playing children, relaxed senior citizens, and loving couples everywhere. They play sports, exercise, chat, and share their joy with friends and family... A brand new lifestyle — "Park lifestyle" has become a part of the locals' daily lives.

(5) Huaqiao Park

The Huaqiao Park, or literally Overseas Chinese Park, is located in Fuzhou, a famous city as the hometown for numerous overseas Chinese. Taking the culture of overseas Chinese as a theme, this project aims at creating various scenarios as a window on the culture of overseas Chinese, and commemorating their contributions to the local development and civilization all over the world. The embossment of a world map acts an iconic landscape at the entrance of the park, representing the distribution of the global overseas Chinese with the 50 national-flag pillars on the grassland. The embossment extends and forms a helm-shaped clock which symbolizes overseas Chinese's spirit of constantly striving for progress. Leaf-shaped green islands as an element connecting the whole park are scattered on the grassland, telling a story of falling leaves settling on the roots and implying the culture of the overseas Chinese returning to their mother country.

The design based on the local cultural characteristics emphasizes on ecological creation. The thriving woods and trails provide a leisure place in the busy urban area to relax and connect with the nature.

项目名称 _ 福州南江滨公园
委托单位 _ 融侨（福州）置业集团有限公司
设计内容 _ 方案及施工图设计
项目面积 _ 约200 000m²
设计时间 _ 2009~2010年

Name _ Fuzhou Nanjiangbin Park
Client _ Rongqiao (Fuzhou) Real Estate Group Co. Ltd
Service _ Landscape design and construction drawing
Size _ about 200 000 m²
Year of design _ 2009~2010

10. 黄昏时分的河滨景致
11. 夜晚的江边是休闲、散步的好去处

10. The riverfront view in dusk
11. At night the riverfront trail becomes a good place for people to have recreational activities, or take a walk.

福州淮安山体公园
HUAI'AN MOUNTAIN PARK IN FUZHOU

山体公园位于福州淮安半岛鬼洞山,为金辉淮安别墅群所环绕,区位优势明显。公园以原有自然山体地形为基础,增设漫步登山道、观景休息平台、森林剧场等配套场所及服务设施,共同组成生态环境健康,可登高眺远、休闲度假的空中花园式公园。位于公园最高处的灯塔更是整个项目的亮点,其不仅是一处景观标志,也是所有登山游客的方向与精神指引。

1. 观景平台

1. View platform

2. 山体公园与金辉淮安别墅群紧紧相连
3. 山体公园登山道平面图
4,5. 蜿蜒的登山栈道

2. The park is surrounded by the Jinhui Huai'an villa clusters.
3. Hiking trails map
4,5. Hiking boardwalk path

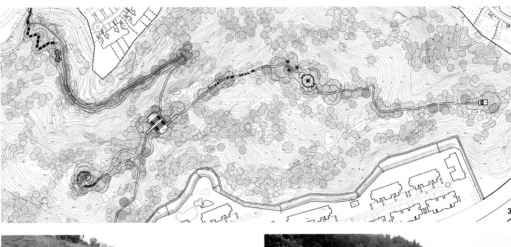

Huai'an Mountain Park is located at the Guidong Mountain in the Huai'an peninsula in Fuzhou, surrounded by the Jinhui Huai'an villa clusters that give the site favorable natural and cultural conditions. Based on the original hilly terrain, hiking trails, view platforms, forest amphitheatres and other amenities and features were designed to form a pleasant, ecological environment, providing visitors a heavenly garden-style park for mountain climbing, overlooking, and leisure activities. The lighthouse at the peak of the mountain is one of the main features in the park, not only as an iconic landscape but also a physical and spiritual guide to all hiking tourists.

项目名称 _ 福州淮安山体公园
委托单位 _ 融侨（福州）置业有限公司
设计内容 _ 方案及施工图设计
项目面积 _ 197 849m²
设计时间 _ 2012年

Name _ Huai'an Mountain Park in Fuzhou
Client _ Rongqiao (Fuzhou) Real Estate Co., Ltd.
Service _ Landscape design and construction drawing
Size _ 197 849 m²
Year of design _ 2012

成都河畔新世界滨河公园
CHENGDU NEW WORLD RIVERFRONT PARK

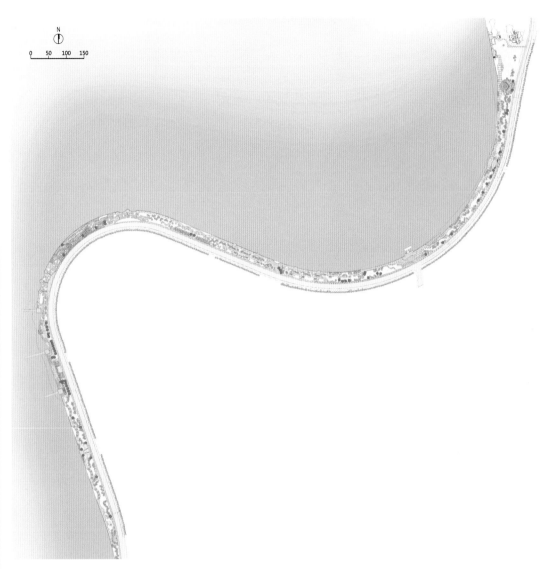

1. 鸟瞰图
2. 总平面图

1. Bird's eye view
2. Site plan

　　河畔新世界滨河公园作为河畔新世界楼盘景观设计项目，地处成都天府新区核心地带，位于成都市中轴线天府大道起点之上。基地呈不规则带状，长约3 000m、宽约50m。基地现状以河滩地为主，植被长势良好，生态结构较为完整；候鸟成群，两栖类动物活动频繁。设计最大限度地对现有场地进行了保留，并通过最小干预将其中部分场地改造为兼具艺术魅力与生态功能的滨河公园。这一滨河地带不仅保留并修复了白鹭栖息地与生态湿地，还添设了儿童乐园（包括风筝草坪、河边嬉戏区、滑板场等）、圆形大草坪、庆典广场等空间，可供居民与游客赏景、游玩、散步、阅读、购物和集会等。这一新的滨河公园将成为当地的"绿色"地标。

公共　Public Space

3. 绿树环绕中的河畔木栈道
4. 植被、河流与栈道组成了宁静、优美的画面

3. Boardwalk along the riverfront
4. Plants, river, and boardwalk form a picturesque scenery.

As the landscape design project for the New World Riverfront residential area, the New World Riverfront Park is located in the center of Tianfu New District at the beginning section of the Tianfu Avenue of Chengdu. Roughly in a belt shape, the site is approximately 3 000 m in length and 50 m in width. Mainly consisting of flood lands and foreshores and with lush plants and rich ecological structure, the site provides habitats for migratory birds and amphibians. The design preserves the existing natural environment as much as possible, and through minimum intervention, transforms part of the site into an attractive riverfront park with multiple ecological functions.

The project not only remains and restores the habitats for egrets and ecological wetlands, but also creates a place for recreational activities such as touring, walking, reading, shopping, and gathering by adding children playgrounds (including kite-flying lawn, riverside play field, and skateboard area), grand lawn, and celebration square. The new park will become a "green" landmark for local district and the city.

5. 水景
6. 蜿蜒的小径通向隐于树木中的建筑
7,9. 驳岸的多种处理方式
8. 清澈的池水中树木的倒影清晰可见

5. Water feature
6. Winding path leading to buildings hidden in the thriving w
7,9. Various design approaches on the revetment spaces
8. Reflecting pond

项目名称	成都河畔新世界滨河公园
委托单位	成都心怡房地产开发有限公司
设计内容	方案及施工图设计
项目面积	72 000m²
设计时间	2009年

Name _ Chengdu New World Riverfront Park
Client _ Chengdu Xinyi Real Estate Co.,Ltd.
Service _ Landscape design and construction drawing
Size _ 72 000 m²
Year of design _ 2009

公共　Public Space

人居　　RESIDENTIAL LANDSCAPE

福州融侨外滩
RONGQIAO WATERFRONT NEIGHBORHOOD IN FUZHOU

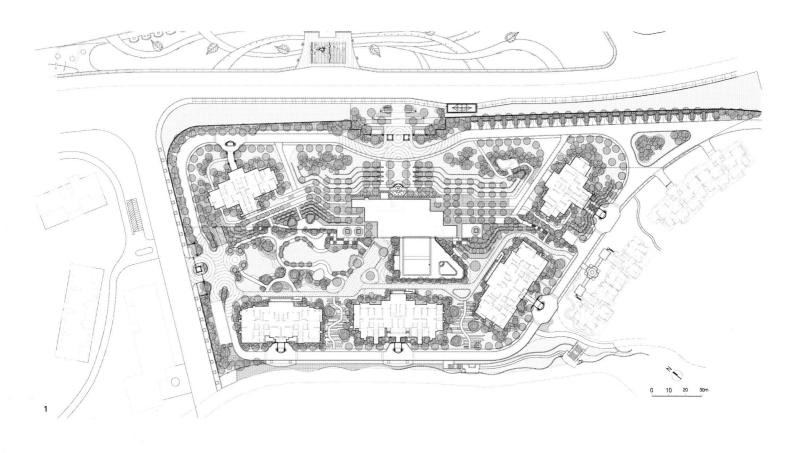

2009~2010年间，融侨集团与景虎研发团队通力合作，针对Art Deco（装饰艺术）风格对一系列地产景观进行了调研与分析，并综合考虑了融侨集团对自身地产景观的需求：福州富裕阶层渴望有所提升的品质生活，并体现出"经典永恒"的时代传承。

为配合场地中Art Deco风格的建筑语汇，设计师尝试将这种风格延伸至景观设计之中，以达到建筑与景观相互呼应的效果。融侨外滩的景观设计在整体上采用现代简洁的大尺度线条来表现台地、主轴的空间效果；同时对经典Art Deco元素进行提炼和总结，融合福州的地域文化特色，创造出新的Art Deco景观语汇，并将这种语汇在小区主入口、车行入口、雕塑、小品、地面铺装等要素中反复应用，从而在视觉上形成了强烈的统一感。此外，在细部的装饰上也沿用了传统Art Deco风格的装饰手法。植物景观的营造结合地形，并根据园区功能规划布局，分为结构性种植、花境类种植和场景化种植三大类，为园区打造出了舒适、简洁、自然的绿色环境。

通过将经典与现代融合，景虎为住户营造出了丰富、典雅、意蕴悠长且极富尊贵感的居所环境。

1. 总平面图
2. 现代而简洁的线条

1. Site plan
2. Modern linear elements

3. 入口景观
4. 融侨外滩与华侨公园无缝连接
5. 入口鸟瞰

3. Main entrance
4. The Rongqiao Waterfront neighborhood is adjacent to the Huaqiao Park.
5. Bird's eye view of the main entrance area

From 2009 to 2010, Rongqiao Group had collaborated with Landhoo research and design team to conduct investigation and analysis of Art Deco style residential landscapes to identify the position and philosophy of the real estate landscapes developed by Rongqiao Group — to create high quality living environment for the wealthy class in Fuzhou and to show the glory of classic design.

To integrate with the Art Deco architectural style of the existing buildings on the site, landscape architects adopted the same style in the landscape design. Large-scale, modern linear elements were employed to stress the unique spatial layout consisting of terraces and landscape axes. Classic Art Deco elements were refined and interwoven with Fuzhou's local characteristics to form new landscape Art Deco elements which were repeated in the design of the entrances, sculptures, landscape features, and pavement, creating a

6. 草阶与道路用以分隔空间
6. Various spaces are divided by the grass terraces and paths.

人居　Residential Landscape

7. Art Deco风格的水景墙
8. 剖面图
9. 每一线条的起点都是一个源于"寿山石印章"的石狮雕塑
10. 入口景观立面图

7. Art Deco style water feature
8. Section
9. Stone lion sculptures as the beginning symbol of each path
10. Elevation of the main entrance

strong identity for the neighborhood. The Art Deco style was also applied in the detailed decorations. According to the existing terrain, three types of planting design were proposed, including structural planting, flower border planting, and feature planting, providing a comfortable, naturalized living environment for the residents.

Through interlacing classic elements and contemporary characteristics, a diverse, elegant, pleasant neighborhood with high identity was built.

人居 Residential Landscape

11. 灯饰、图样、材质共同体现出了Art Deco风格
12. 多层次的使用空间
13. 草阶

11. Lighting, decoration patterns and materials embody the Art Deco style.
12. Multi-level spaces
13. Grass terraces

14. 硬景与软景的融合、渗透
15. 滨水休闲平台
16. 沙坑、滑梯、木马等游戏设施丰富了儿童活动场地

14. Interwoven hard- and soft-scape
15. Waterfront view platform
16. Children playground consists of sandpit, slide, hobbyhorse, and other facilities.

人居　Residential Landscape

17. 种植设计也充满了线条感
18. 汀步增加了平行的道路之间的交通联系
19. 草坪成为午后休闲的良好场所

17. Rectilinear elements were applied in planting design.
18. Stepping stone acts as a connection between parallel paths.
19. The lawn becomes a wonderful recreational place.

项目名称 _ 福州融侨外滩
委托单位 _ 融侨（福州）置业集团
设计内容 _ 方案及施工图设计
项目面积 _ 36 750m²
设计时间 _ 2010~2011年

Name _ Rongqiao Waterfront Neighborhood in Fuzhou
Client _ Rongqiao (Fuzhou) Real Estate Group
Service _ Landscape design and construction drawing
Size _ 36 750 m²
Year of design _ 2010 ~ 2011

人居　Residential Landscape

重庆金辉春上南滨·江城著
KAMFEI RIVERFRONT NEIGHBORHOOD IN CHONGQING

1. 道路设计现代、简洁的平面构图
2. 总平面图
3. 小区入口

1. Modern, concise design of the pedestrian circulation
2. Site plan
3. Main entrance

人居 Residential Landscape

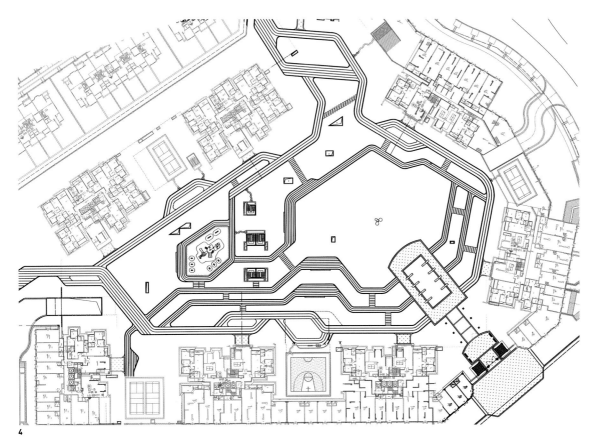

4. 中心区域平面图
5. 宅间小路
6. 宅间局部平面图

4. Detailed plan of the central area
5. Beautiful landscape along the paths
6. Enlarged plan

项目位于重庆南岸区西部，北临长江，拥有长江半环绕的优质资源和4km长的怡人江岸线。同时场地邻近铜滨公园、融侨公园等城市绿地，环境怡人。项目用地呈不规则三角形，基地南高北低，高差约22.8m。项目业态由高层住宅、别墅和商业构成。

建筑设计汲取Art Deco风格精华：采用对称的构图、刚柔并济的线条、形式多样的浮雕、强调质感与光泽的材料；同时采用鲜艳大胆的纯色、对比色和金属色，追求强烈、华美的视觉效果。

景观设计中延续了福州融侨外滩项目的风格特色：平面构图运用现代简约的设计手法，通过图案感极强的平行线条划分园区内的各个空间，提供优雅、简约、现代景致的同时，解决了园区内的功能需求问题，真正实现了功能与美学的完美统一。富有韵律感的主入口和台地设计、大片开阔的平坦草坪和成列栽植的树木都体现了统一和均衡的设计原则。在平面构图中，每户单元的出入口都形成了一个交通的转换点，形同两只相互交握的手，这一设计也体现了金辉集团追求创新与品质的企业理想——通过和谐景致的营造为客户提供超越预期的生活品质。

Located in the west part of Nan'an District, Chongqing, to the south of Yangtze River, the site has rich, quality potential of being riverfront. At the same time, the nearby Tongbin Park and Rongqiao Park provide a pleasant urban green circumstances for the project. In a roughly triangle shape, the site is high in the south and low in the north with a height difference of 22.8 m, consisting of high-rise housing area, villa area, business areas and other urban developments.

Absorbing the essence of Art Deco style, the architectural design adopted a symmetric composition, with rectilinear and winding linear elements, and embossments in diverse styles; texture and luster were well considered in the selection of structural materials. A strong, spectacular effect was achieved by using bold solid colors and creating sharp visual contrasts.

The landscape design followed the Art Deco style employed in Fuzhou Rongqiao Waterfront project: different spaces were formed by parallel linear elements, such as paths and green belts, creating a comfortable living environment while organizing the both vehicle and pedestrian circulations. The rhythmical layout consisting of the main entrance, terraces, open grassland, and lines of trees has embodied a sense of harmonious beauty, aesthetically and functionally. In particular, the entrance of each unit forms a turning point, just looking like two holding hands, which implies the corporate philosophy of Kamfei Group, that is to establish a quality life-style by offering an unexpected, exquisite living environment.

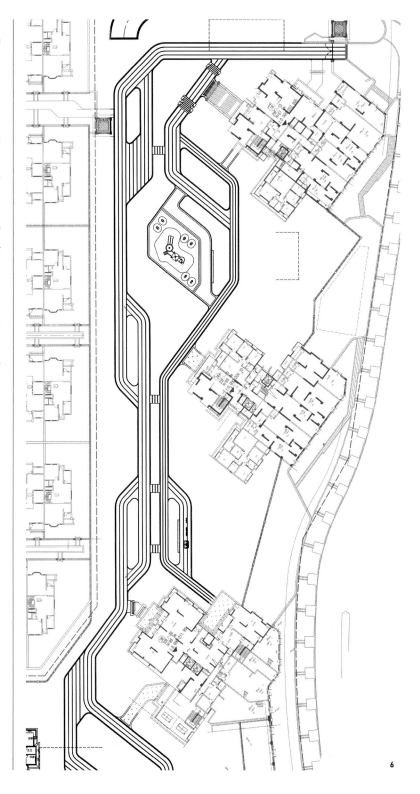

6

项目名称 _ 重庆金辉春上南滨·江城著
委托单位 _ 金辉集团融侨长江（重庆）房地产有限公司
设计内容 _ 方案及施工图设计
项目面积 _ 96 200m²
设计时间 _ 2013年

Name _ Kamfei Riverfront Neighborhood in Chongqing
Client _ Kamfei Group Rongqiao Changjiang (Chongqing) Real Estate Co., Ltd.
Service _ Landscape design and construction drawing
Size _ 96 200 ㎡
Year of design _ 2013

成都华韵天府·三径
CHENGDU HUAYUN TIANFU

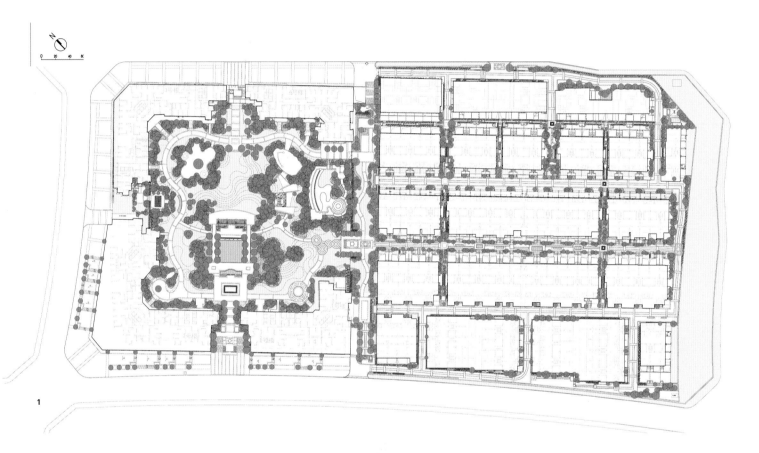

1

"濯锦之江，源远流长"，成都市锦江区自古人文荟萃，底蕴厚重。自唐宋以来，便因"百业云集，市廛兴盛"而成为饮誉川西的繁华贸易区。此外，商业文化、宗教文化、民俗文化、酿酒文化、藏书印刷文化、蜀锦文化、陶瓷文化等多种文化单元融合而成的锦江文化，其内涵丰富、形式多样。

如今，清洁、绿色、生态、健康理念已融入锦江发展。锦江区内的国家AAAA级景区"三圣花乡"，花卉种植历史源远流长，是著名的"中国花木之乡"，已成为市民观光娱乐、休闲度假的绝佳去处。据碑文记载，这里有座建于清代的三圣庙，供奉炎帝神农氏、黄帝轩辕氏、黄帝史官仓颉；后改奉刘备、关羽、张飞，因而得名"三圣乡"。这一区域是成都市城市东部副中心——"东部新城"的起步区，科研院所云集、文化氛围浓郁。

华韵天府项目地处"东部新城"起步区的中轴地段，临近锦江区的行政中心。场地占地面积40 333m²，为典型的天然浅丘坡地风貌，奠定了营造高品质景观的基础。

该项目分为两期开发，一期为户型面积70~122m²的高层住宅楼，二期为户型面积230~350m²的三进三层传统中式巷院别墅。在景观设计中，我们结合建筑风格运用现代中式的设计手法，以"花"文化为背景，打造具有传统特色的西式民居空间。设计取意国画山水，营造出丰富多姿的花境，并以季节性观花乔木作为视觉亮点，力图创造"花重锦官城"的意境。

巧妙的空间布局：首先从对项目的整体性和景观的延续性考虑出发，在高层区与别墅区之间用"街巷"作为连接通道。街巷以及院墙曲折蜿蜒，产生出富于变化的景观节点和空间层次，以丰富的

1. 总平面图
2. 鸟瞰图
3. 入口标志

1. Master plan
2. Bird's eye view
3. Main entrance

人居　Residential Landscape

植物作为背景，形成川西民居中古街巷韵味的风景。

经典的主题元素：运用现代中式的设计手法，将花的具象性特征演绎为景观符号，在景观水体、地面铺装、景观构筑物、雕塑小品等要素中，将"花"文化贯穿始终。

诗情花意的命名：项目高层区的4栋高层住宅共有10个单元，分别以中国十大传统名花命名，并根据古人咏十大名花的名句分别题名为：云裳（牡丹）、逊雪（梅花）、幽芳（兰花）、谪仙（荷花）、东篱（菊花）、雪娇（山茶）、长春（月季）、岁红（杜鹃）、月影（桂花）、玉骨（水仙）。

4. 对称式的景观布局,加入传统古街古巷文化元素,体现出浓郁的川西民居特色
5. 采用传统园林手法展现现代中式景观

4. A western Sichuan dwelling style was created by applying symmetrical landscape layout and ancient alley elements.
5. Neo-Chinese landscape created through traditional gardening approaches.

Jinjiang River runs a long history. Jinjiang District in Chengdu has rich cultural and humanity heritages. It has been a prosperous commercial and trading center in the western Sichuan Basin since the Tang Dynasty. The Jinjiang culture is characterized as a fusion of business culture, Taoist and Buddhist cultures, folklore culture, liquor-making culture, book collection and printing culture, Sichuan brocade culture, and ceramic culture.

In recent years, the concepts of cleanness, green, ecology and health have been integrated into the development of Jinjiang District. The Three Saints Flower Town, a national AAAA-level scenic spot, has become public sightseeing and recreational destination. The name of Three Saints Town

6. 别墅区入口景墙
7. 景墙立面图
8. 川西休闲茶馆文化
9,11. 院落划分形成的窄巷
10. 利用辅首作为景墙装饰元素

6. Landscape wall at the entrance of the villa area
7. Elevation of the landscape wall
8. Western Sichuan tea-house
9,11. Laneway between courtyards
10. Traditional Chinese door knocker as decoration element

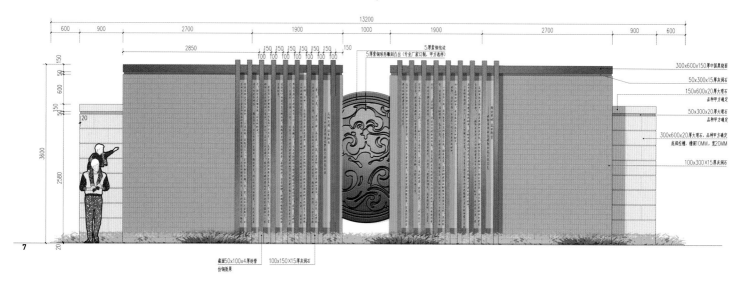

人居　Residential Landscape

stems from the tablet inscription that records there was a memorial hall named Three Saints Temple which was built in the Qing Dynasty to worship the Emperor Shen Nong, Emperor Xuan Yuan and the Great Historiographers Cang Jie, and was changed to worship Liu Bei, Guan Yu, and Zhang Fei, three famous heroes in the Three Kingdoms Age. This area now is a sub-center of the eastern new development area of Chengdu, as one of the first development parcel of "East New Town", housing various research institutes. The Three Saints Town has a long history of flower planting, known as the famous "hometown of flower in China".

The project is located at the axis of the starting area of the East New Town, near the administrative center of Jinjiang District. The project covers an area of 40 333 m^2, and the site is characterized as a typical natural shallow hilly area, providing ideal terrain conditions for creating a high quality landscape.

The project comprises two main phases: phase I development for high-rises with apartment area of 70 ~ 122m^2; phase II development for traditional Chinese courtyard villas with floor area of 230 ~ 350m^2. In the landscape design neo-Chinese style was applied, which responded to the modern architectural style of the existing buildings on the site; taking the flower culture as its context, a traditional western Sichuan-style residential space was created. The design drew the inspiration from Chinese painting. Under the theme of flower, various flower borders were

created and seasonal blossoming trees were planted as focal features forming an ambience of "blossoming paradise".

Ingenious special layout: First, based on the consideration on the whole project and the continuity of the landscape design, "alleys" were designed to connect the high-rises and the villas. Winding streets and walls form changing landscape and multiple spatial layers, creating rustic, ancient, western Sichuan-characterized alley scenery through using diverse plant materials.

Classic theme elements: Using neo-Chinese design

approaches, flowers were figuratively translated as landscape elements and symbols, and interwoven into the waterscape, pavement, structures, sculptures and other features.

Poetic naming after flower implication: There are 10 units in the four high-rises, which are named after famous poem about ten greatest traditional Chinese flowers: Yunshang (peony), Xunxue (plum blossom), Youfang (orchid), Zhexian (lotus), Dongli (chrysanthemum), Xuejiao (camellia), Changchun (Chinese rose), Suihong (azalea), Yueying (sweet osmanthus) and Yugu (narcissus), respectively.

12. 巷院间的节点空间
13. 别墅内的休息空间鸟瞰
14. 地下一层处理为可供使用的天井空间
15. 特色的景观"门"
16. 巷院内的水钵
17. 设计再现了川西民居古街古巷的情景

12. The simple, elegant design of the junction of alleys
13. Bird's eye view of the courtyard
14. The basement of each villa can be utilized as an intimate courtyard
15. Decorated access
16. Water vat
17. Ancient charm of western Sichuan dwellings

18. 典雅的别墅入口景观
19. 小型的院落空间
20. 幽然小径
21. 转折处的空间变化

18. Refined, elegant landscape design of the entrance of each villa
19. Small yard between the villas
20. Tranquil path
21. Detailed design of turning

项目名称 _ 成都华韵天府·三径
委托单位 _ 成都金房营销策划有限公司
设计内容 _ 方案及施工图设计
项目面积 _ 40 333m²
设计时间 _ 2011年

Project name _ Chengdu Huayun Tianfu
Client _ Chengdu Jinfang Marketing Planning Co., Ltd.
Service _ Landscape design and construction drawing
Size _ 40 333 m²
Year of design _ 2011

南充水慕清华

NANCHONG SHUIMU QINGHUA COURTYARD HOUSES

1. 入口景观
2. 竹与白墙

1. Landscape design of the main entrance
2. Bamboos and white walls

南充市是"三国文化"的源头（《三国志》作者陈寿的故乡），也是古代天文研究中心（西汉著名天文学家落下闳的故乡），因此设计师将这些地域文化内蕴注入项目的设计方案之中。水慕清华项目定位为现代中式住宅，设计师在整体设计中运用新中式手法，对传统住宅进行演绎。项目整体给人一种古朴、典雅又不失现代的亲和感。

新中式特色

千百年来中式民居外立面大多是以"黑、白、灰"三种色彩

为主,"素"是中式民居的主要特色。古朴的白墙灰瓦、茂密的竹林、镂空的砖墙、方圆结合的局部造型、青石铺就的小巷、半开放式的庭院、通透的漏窗、文化牌坊等要素都在该项目中得到体现。

种植特色

设计师尝试将建筑白墙灰瓦的淡雅与绿色植物的清新相结合,在住宅边、花窗后、小径旁、转角处等区域种植了竹林。人们可以透过竹林隐隐约约地看到墙壁,缓解了大面积素色的单调、压抑之感。除此之外,设计师还在步道系统中种植了一些姿色俱佳的乔灌木,以达到遮阴、纳凉的效果;再配以芭蕉等具有浓郁文化韵味的植物,使得窄街深巷、高墙小院显得更为深邃清幽。

功能设置

设计师在该项目中设置了不同功能的场所,如儿童活动天地、羽毛球场地、休闲品茗场地和健身场地等。

Nanchong City is the cradle of the culture of Three Kingdoms (since Nanchong was the birthplace of Chen Shou, the author of *History of the Three Kingdoms*), and the ancient center of astronomical study (since Nanchong was the hometown of Luo Xiahong, a famous astronomer in Western Han Dynasty). In the Shuimu Qinghua project, landscape architects integrated the local cultural context with the design to build a neo-Chinese style residential area. By interpreting traditional Chinese houses with modern landscape approaches, the project creates a harmonious ambiance that is primitively simple, elegant and modern.

Neo-Chinese Characteristics

For thousands of years, the colors of black, white and gray have been widely used in traditional Chinese houses, establishing a distinctive residential style that is unadorned but elegant. Elements such like simple white walls with gray tiles, clustered bamboos, hollowed-out brick walls, rectangle- and circle-shaped decorations, alleys with bluestone paving, semi-open courtyard, refined traditional latticed windows, and memorial archways can all be found in this project.

Planting Design

Designers employed the palette that comprises white walls

3. 中式园林中常用门洞来分隔空间
4. 白墙、灰瓦、绿竹形成一幅优美画面
5. 整打青石的鱼池
6. 拴马桩石雕

3. Doors are used to separate spaces in Chinese gardens.
4. White walls, gray tiles, and green bamboos form a picturesque scenery.
5. Small fish pond made of one single piece of bluestone
6. Sculptures

and gray tiles of the architecture with the pure and fresh green of plants. Bamboos were planted next to the houses, behind the latticed windows, by the paths and around the corners. Through the bamboos walls can be vaguely seen, being a relief of a sense of monotony and depression caused by the large-area plain colors. In addition, a variety of trees and shrubs were planted in the pedestrian environment for shadows; plants with strong cultural charm, such as plantains, were also planted to bring profoundness and tranquility to the narrow alleys, high walls, and small courtyards.

Functional Spaces

Places with different functions were created, including children playground, badminton courts, tea houses and fitness centers.

项目名称 _ 南充水慕清华
委托单位 _ 南充川北鑫益房地产开发有限责任公司
设计内容 _ 方案及施工图设计
项目面积 _ 32 586m²
设计/建成日期 _ 2006~2008年

Name _ Nanchong Shuimu Qinghua Courtyard houses
Client _ Nanchong Northern Sichuan Xinyi Real Estate Development Co., Ltd.
Service _ Landscape design and construction drawing
Size _ 32 586 m²
Year of design/completion _ 2006 ~ 2008

绵阳长虹南山一号
CHANG HONG NO.1 MANSION IN NANSHAN, MIANYANG

南山一号项目位于绵阳市涪城区南山，此处三江交汇，涪城美景尽收眼底，自然景观上佳。就好比需要用巧工才能雕琢好玉，如此优越的景观条件，更是对设计师"手艺"的考验。设计师结合绵阳的历史文化，以及场地在城市中的重要地位，确定了其主题概念与设计思路——带有装饰韵味的现代中式风格，并以传统文化中的"六艺"（礼、乐、射、御、书、数）作为主题定位。六大主题景观区如同彩虹一般呈现不同的色彩、不同的韵律，相辅相成且生动活泼。

"礼"主题景观区：该区位于社区入口处，打造了拥有仪式感的入口形象空间，体现了"礼"的主题。包括特色瓦罐列置（体现队列的仪式感）、酒樽（古代欢迎来客的酒礼）、五礼亭（五礼指古代的五种礼制）和周礼景墙（《周礼》为儒家经典著作）。

"乐"主题景观区：借鉴古典园林的"一池三山"造园手法，以枯山水和疏林草地为主要元素打造"高山流水"（经典古曲）景观，潺潺水流声亦可令游人得到听觉享受。

"射"主题景观区：与射箭文化相结合，打造雕塑草坪景观；场地高差采用文化墙来处理，营造垂直景观效果，并设计有主题剪影墙和雕塑。

"御"主题景观区：结合休闲广场、地面浮雕（马车的文

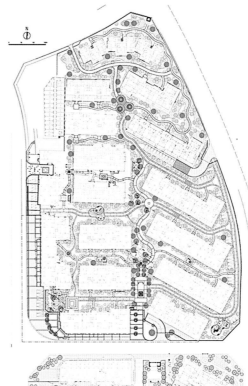

1. 售楼部前广场
2. 总平面图
3. 重复出现的拴马桩等中式文化符号

1. Square in front of the selling center
2. Site plan
3. Chinese cultural symbols such as ancient hitching post are repeated in this project.

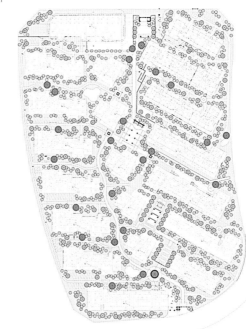

化记忆），打造多功能的休闲节点；利用车马图的景墙来诠释"御"的主题；植物采用自然式处理手法，打造舒适温馨的休闲环境。

"书"主题景观区：在售楼部前广场展示书的文化，具体手法包括，构建围合空间，折射出书苑文化；用地面浮雕对书画进行诠释；并设计了能体现诗词文化的景墙，烘托出场地的文化氛围。

"数"主题景观区：将数字趣味游戏融入石阵景观中，体现"数"的主题；结合《九章算术》设计了数字石阵，为儿童与青少年创造了益智活动空间。

人居 Residential Landscape

The project is located at Nanshan in Fucheng District, Mianyang City where three rivers meet with beautiful natural landscape views. Just like that a top-grade jade art piece requires delicate craftsmanship, a place with outstanding landscape condition requires delicate design. Have considered on the local history and culture, as well as the geographical importance in the city, designers identified the core concept and tone of the plan — expressing "Six Traditional Arts" in modern Chinese style with ornamental elements. The six landscape zones, which respectively feature the six arts were designed in different colors and patterns, complementing each other.

Zone of "Rites" theme: This area is located near the entrance of the community to create a ritual-sense space and highlight the theme of "Rites". Specific approaches and elements include pottery array, landscape featured with alcohol goblets elements (ancient etiquette to welcome guests), Pavilion of Five Etiquette (five ancient etiquette systems), and landscape wall with Zhou Li elements (*Zhou Li*, the classic Confucius master piece).

Zone of "Music" theme: Drawing inspiration from the "one pool three hills technique" commonly used in the traditional Chinese gardens, the landscape of Mountain and Stream (classical

Chinese tune) was created with landscape featured Karesansui and open grassland where visitors can enjoy the rippling sound of streams.

Zone of "Archery" theme: Integrated with the archery culture, the sculpture lawn was created. A landscape wall was designed by taking advantage of the elevation change, while using culture wall to form a vertical landscape feature with silhouette and sculptures.

Zone of "Chariot driving" theme: Multi-functional recreational spaces were designed combining with squares and ground embossment (to evoke the history of carriage culture). The carriage-image wall was designed to respond to the "chariot driving" theme. The plants were naturally arranged to create a pleasant, attractive recreational environment.

Zone of "Calligraphy" theme: Calligraphy culture was embodied through the pavement and landscape wall in front of the selling center, including the enclosed space (academy of classical learning), ground embossment (calligraphy and Chinese paintings), and culture landscape wall (classical Chinese poetry).

Zone of "Mathematics" theme: This zone integrates maths games with stone forest to show the charm of ancient mathematics. The stone forest designed based on *Nine Chapters on Mathematical Procedures* provides recreational and educational space for kids and teenagers.

4. 露天休息空间
5. 蜿蜒的园路
6. 室内装饰呈现中国风韵

4. Outdoor rest space
5. Winding path
6. Traditional chinese style interior decoration

项目名称 _ 绵阳长虹南山一号
委托单位 _ 四川长虹电子集团有限公司
设计内容 _ 方案及施工图设计
项目面积 _ 65 475m²
设计时间 _ 2012年

Name _ Chang hong No.1 Mansion in Nanshan, Mianyang
Client _ Sichuan Changhong Electronics Group Co., Ltd.
Service _ Landscape design and construction drawing
Size _ 65 475 m²
Year of design _ 2012

成都置信芙蓉古城·紫微园

ZHIXIN EMPEROR STAR NEIGHBORHOOD IN FURONG ANCIENT TOWN AREA, CHENGDU

1. 全景鸟瞰
1. Bird's eye view

"紫微园"坐落于自然条件得天独厚的成都西郊国家级生态示范区——温江永宁镇，是置信芙蓉古城开发收官之作。建筑采用中国传统的宅院、院落风格与现代材料相结合的方式，打造精品高档别墅区。项目景观设计借鉴皇家园林及川西、江南传统园林的设计手法，厚载人文，融古典于时尚，扬传统之精粹。山、水、石、桥、亭、榭、廊等相互穿插交融，浑然天成，造就独一无二、清逸静雅的理想居住氛围。特别是为每户住宅设计的专属庭院水景，更是集传统园林之大成，并加入当地特色植物搭配，彰显尊荣华贵。

As the last work developed in the Furong Ancient Town Area by Zhixin Development, the project is located in the Yongning Town in Wenjiang which is a beautiful national ecological demonstration area in the western suburbs of Chengdu. Traditional Chinese structures including houses and courtyards were adopted in the architectural design, combining with modern construction materials to create elegant villas. Drawing lessons from the design approaches used in traditional Chinese royal and private gardens, the landscape design was integrated with the rich local cultural heritage, forming a neo-classical style. Mounds, water bodies, stone settings, bridges, pavilions, and corridors were delicately arranged, creating an ideal living environment of uniqueness and elegance. In particular, the private courtyard with water features designed for each house was decorated with various native plants, providing distinctive, glorious private spaces.

2. 内庭景观
3. 总平面图
4. 沿水景布设的休憩亭廊与散步道

2. Landscape of the inner garden
3. Site plan
4. Pavilions and corridors along the water features

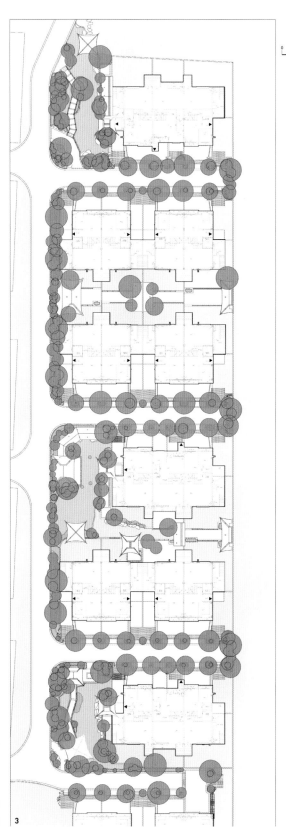

项目名称 _ 成都置信芙蓉古城·紫微园	**Name** _ Zhixin Emperor Star Neighborhood in Furong Ancient Town Area, Chengdu
委托单位 _ 成都置信房地产开发有限公司	**Client** _ Chengdu Zhixin Real Estate Development Co., Ltd.
设计内容 _ 方案及施工图设计	**Service** _ Landscape design and construction drawing
项目面积 _ 3 185m²	**Size** _ 3 185 m²
设计时间 _ 2013年	**Year of design** _ 2013

人居　Residential Landscape

成都青城郡
QINGCHENG VILLA IN CHENGDU

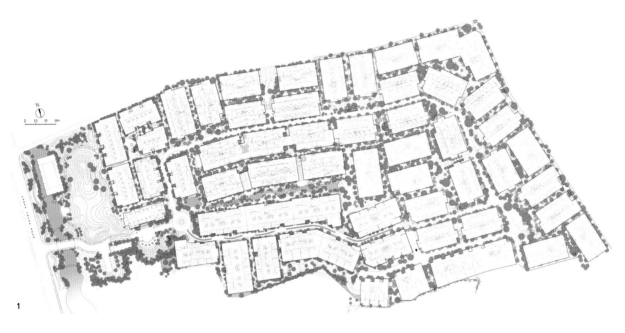

项目位于都江堰青城山环山景观带，是青城山区域中富有代表性的英式小镇风情别墅群。该项目以自然野趣作为景观主题定位，通过英式自然草坪、小桥流水，以及温馨的私家花园等景观，营造典雅、静谧的高品质生活氛围。

1. 总平面图
2. 公共区域景观
3. 休闲大草坪

1. Site plan
2. Landscape of public area
3. Lawn

人居　Residential Landscape

The project is located in the Dujiangyan City as a part of the landscape belt of Qingcheng Mountain. It is a villa cluster of typical British style and takes charming, natural landscape as its design theme. Through choreographing the elements including British-style natural lawn, bridges, as well as private gardens, an elegant, tranquil, quality living environment was created.

4. 草坪空间中的自然狩猎题材雕塑
5. 院落一隅的休闲空间
6. 别墅外的小径

4. The hunting-theme sculptures on the lawn
5. Recreational space
6. Path connecting the villas

人居　Residential Landscape

7. 古朴的路灯和特色景观小品——电话亭
8. 水车与小屋

7. Landscape furniture — street lamp and phone booth
8. Waterwheel and cabin

人居　Residential Landscape

9. 乔灌木巧妙配植，营造出色彩丰富的园路景观
10. 别墅入口景观

9. A colorful, stunning streetscape was created through exquisite plant design.
10. Landscape of the front door of each villa

项目名称 _ 成都青城郡	**Name** _ Qingcheng Villa in Chengdu
委托单位 _ 四川金联实业有限公司	**Client** _ Sichuan Jinlian industry Co., Ltd.
设计内容 _ 方案及施工图设计	**Service** _ Landscape design and construction drawing
项目面积 _ 73 600m²	**Size** _ 73 600 m²
设计时间 _ 2012年	**Year of design** _ 2012

南京融侨观邸
RONGQIAO MANSION IN NANJING

　　融侨观邸地处南京浦口新城和主城区的交汇处——珠江镇。珠江镇与主城区的奥体中心隔江相望，是未来江北副城市中心的核心。该项目占地面积约28万m²，总建筑面积48万m²。

　　该项目建筑设计定位为法式风格，景观设计也延续了这一格调。设计强化了小空间"小"的特色，通过层次丰富的垂直绿化、严谨细致的景观细节——尤其是芳香四溢的香草丛林，营造出具有浪漫气息的景观氛围；设计也突显了大空间"大"的特色，以疏林草地的形式，明确限定了空间范围，并通过景观序列的组织，达到步移景异的景观效果；与此同时，设计运用园区内的自然坡度以形成山丘地形，创造出高低错落、层次丰富的景观效果。最后将诸多法式园林（如凡尔赛庭院、枫丹白露宫等）中的经典元素融合在设计之中，充分展现了风雅的法式风格。

Rongqiao Mansion is located in Zhujiang Town, the junction of the Pukou new district and the existing city center of Nanjing. Zhujiang Town faces the Olympic Sports Center across the Yangtze River, and is considered as the future sub-city center in the north area of the city. Covering an area of approximately 280 000 m^2, the site has a total building area of 480 000 m^2.

French style is not only adopted in the architectural design, but also in the landscape proposal. The design emphasizes the "elaborate" characteristic of the small-scale spaces by creating romantic atmosphere through richly-layered vertical vegetation and well-organized landscape features — especially the woods of aromatic plants. The "grand" characteristic of large open spaces is also highlighted through the landscape form of parklands that defines the spatial layout and creates changing scenery for the site. At the same time, the natural slope is formed taking advantage of the existing terrain, allowing designers to create various landscape layers and features.

A range of elements abstracted from the most classic French gardens (such as the Versailles Palace and Fontainebleau) are integrated into the design to echo the overall French style.

1. 利用园区内的高差变化形成休息空间
2. 总平面图
3. 精致的花钵组景

1. Rest space is formed based on the elevation difference of the existing terrain
2. Site plan
3. Landscape furniture

人居 Residential Landscape

4. 法式别墅与自然式植物景观
4. French villa and natural-style planting design

5. 静谧的小花园
6. 花园俯瞰
7. 水景
8. 欧式园亭

5. Tranquil private garden
6. Bird's eye view over private garden
7. Water feature
8. Pavilion

9. 轴线景观
10. 被绿意环绕的建筑
11. 与建筑群相接的滨水空间
12. 节点水钵

9. Axial landscape
10. Bird's eye view over private garden
11. Waterfront space in front of the buildings
12. Fountain

项目名称	南京融侨观邸
委托单位	江苏融侨置业有限公司
设计内容	方案深化及施工图设计
项目面积	126 006m²
设计时间	2011年

Name _ Rongqiao Mansion in Nanjing
Client _ Jiangsu Rongqiao Real Estate Co., Ltd.
Service _ Detailed landscape design and construction drawing
Size _ 126 006 m²
Year of design _ 2011

崇州炎华置信·上林西江
CHONGZHOU YANVA SHANGLIN XIJIANG VILLAS

1. 依水而居
2. 庭院水景
3. 建筑被茂盛的植被所掩映

1. Living by water
2. Water body in courtyard
3. Buildings surrounded by flourish plants

  项目位于四川省崇州市崇阳区滨河路，整个社区景观定位为英式格调，形成特有的精神气质。从"色"（建筑颜色与地面铺装、小品、灯柱、雕塑的色彩搭配呼应）、"形"（拥有形式各样的广场、花园、水景等景观节点）、"质"（突显铁艺、石材的质感）三方面来打造景观，并结合丰富水系，形成依水而居的格局。此外，还有精心设计的英式花园，共同营造出尊贵、自然、高雅的岛屿生活。

人居　Residential Landscape

The project is located at the Binhe Road in Chongyang District of Chongzhou City, Sichuan Province. British style was identified as the core concept and theme of landscape design, highlighting three main qualities: the "color" — the color scheme of the buildings matches that of pavements, ornamental structures, lampposts, and sculptures; the "shape" — in variety of landscape nodes including squares, gardens, water features, etc.; and the "quality" — texture of iron and stone. Combining with the layout of the water system, these three qualities were integrated to form diverse residences sitting by water. Together with the classic British gardens, the project creates a noble, leisurely, elegant island lifestyle.

4. 欧式景亭
5. 车行道转弯处的植物景观
6. 总平面图

4. Pavilion
6. Landscape design for the turning point
6. Site plan

项目名称 _ 崇州炎华置信·上林西江	Name _ Chongzhou Yanva Shanglin Xijiang Villas
委托单位 _ 四川武田房地产开发有限公司	Client _ Sichuan Wutian Real Estate Development Co., Ltd
设计内容 _ 方案及施工图设计	Service _ Landscape design and construction drawing
项目面积 _ 30 000m²	Size _ 30 000 m²
设计时间 _ 2011年	Year of design _ 2011

西安天朗蔚蓝花城
XI'AN TIANLANG WEILAN HUACHENG

蔚蓝花城位于陕西省西安市，是未来最具发展潜力的中央居住区。景观设计基于项目原有地块方正、地势平坦、路网通畅等特点，结合场地中原生的名贵树木与清新幽静的环境，在"魅力小城，花样生活"的景观主题下，打造出阳光、运动、健康、休闲的大型生态社区。

1. 曲线构图形成优美的中央水景
2. 中央水池
3. 总平面图

1. Central waterscape in a curve layout
2. Central waterscape
3. Site plan

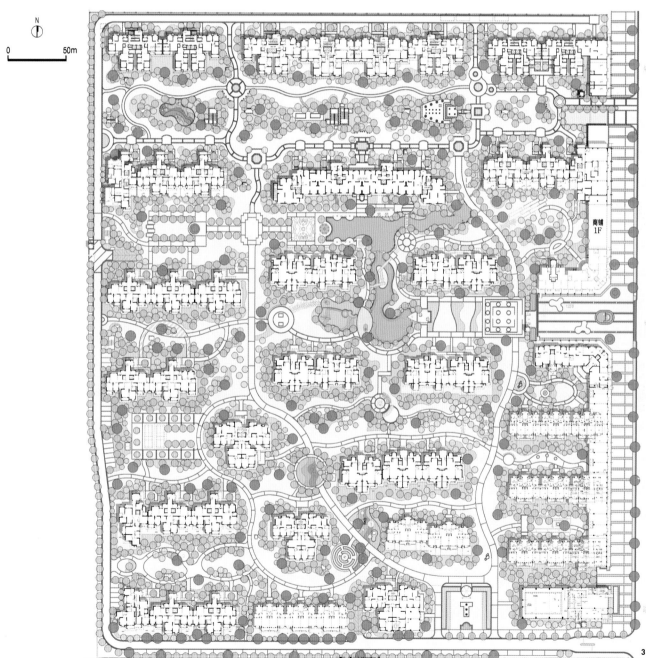

人居　Residential Landscape

The project is located in the future central residential development area of Xi'an, Shaanxi Province. The landscape design took advantages of the existing conditions of the site, including the flat and square terrain, and the convenient transportation network. Integrating with the existing, valuable native trees and the tranquil, refreshing environment, the design created a comfortable, healthy eco-community under the theme of "charming neighborhood, vibrant lifestyle".

4. 公共区域节点空间
5. 欧式景亭
6. 景亭位于模纹花坛构成的中轴线上

4. Landscape design for a public space
5. Pavilion
6. Symmetrical landscape design

项目名称 _ 西安天朗蔚蓝花城	**Name** _ Xi'an Tianlang Weilan Huacheng
委托单位 _ 陕西华宝实业有限公司	**Client** _ Shaanxi Huabao Industrial Co.,Ltd
设计内容 _ 方案及施工图	**Service** _ Landscape design and construction drawing
项目面积 _ 91 650m²	**Size** _ 91 650 m²
设计时间 _ 2010年	**Year of design** _ 2010

西安天朗·大兴郡之锦城

JINCHENG OF TIANLANG DAXING RESIDENTIAL AREA IN XI'AN

1. 总平面图
2. 中轴景观鸟瞰

1. Site plan
2. Bird's eye view of the axial landscape

　　项目位于西安城西大兴东路北侧，环揽汉长安城遗址公园、汉城湖和大兴公园；场地周边还兴建有大汉民俗街、大兴广场及国际商贸基地等项目。整个片区历史文化资源优越，文化氛围浓厚。

　　天朗·大兴郡项目将建筑风格定位为"新汉风"：采用欧式建筑"三段式"的构图法则，并以汉代元素符号为细部装饰，整体呈现中西合璧的建筑风貌。为了使景观与建筑的风格整体一致，设计师构思了一种独具创意的设计手法——将简欧构图法则与具有经典汉代特征的细部元素相结合。在空间构成上，采用简洁的线条，将空间层次划分清晰；在细部装饰上，将经典的汉代图案提炼为文化符号，形成一种全新的汉风Art Deco景观风格。

　　The project is located at the north of East Daxing Road in Xi'an, surrounded by the Heritage Park of Chang'an in Han dynasty, Hancheng Lake and Daxing Park. Other commercial programs such as the Han folklore street, Daxing Plaza and the International Trading Center are under construction. The historical and cultural resources of the site are rich and diverse.

　　The neo-Han style was adopted in the architectural design that combines the ancient Chinese and western architectural essence with the classic European three-section layout and decorative elements of Han dynasty culture. This innovative design language was also employed in the landscape design — spatial layers is divided by clear, geometrical landscape linear elements, and Han symbols are abstracted and reinterpreted in the detail decorations that form a brand new Art Deco landscape design.

3. 丰富的景观元素组合成优美的画卷
4. 楼间草坪

3. Picturesque scenery is created by diverse landscape elements.
4. Open space between buildings

项目名称 _ 西安天朗·大兴郡之锦城
委托单位 _ 西安天朗置业有限公司
设计内容 _ 方案及施工图设计
项目面积 _ 37 000m²
设计时间 _ 2010年

Name _ Jincheng of Tianlang Daxing Residential Area in Xi'an
Client _ Xi'an Tianlang Real Estate Co., Ltd.
Service _ Landscape design and construction drawing
Size _ 37 000 m²
Year of design _ 2010

成都万华·麓山国际社区之麓镇样板区

LU TOWN DEMONSTRATION AREA OF THE WANHUA LUSHAN INTERNATIONAL NEIGHBORHOOD, CHENGDU

1. 缓坡草坪逐渐延伸至滨水平台，蜿蜒的小路穿插其中，形成意境幽远的景观空间
2. 趣味性景观雕塑

1. The waterfront lawn and platform provide a tranquil, beautiful space.
2. Sculpture

项目位于成都市双流县麓山大道，周边配套设施完善，交通便利，拥有极佳的居住环境和文化氛围。该社区采用了国际先进的PUD（计划单元综合开发）模式，并完整地引入了北美及欧洲的经典住宅形态，所有建筑都以低密度的姿态融入麓山的自然环境之中。

本项目最大的特色就是将绿色空间引入社区，通过一系列严谨而丰富的细节，营造出美好、和谐的宜居景观氛围。设计师对社区进行了区域划分，力图强调景观的层次感与秩序感，旨在打造出由全开放空间到半开放空间，再逐步向私密空间过渡的体验流线，实现步移景异的景观效果。设计团队按照自然界植物的生长状况和生态习性进行优化组合，将乔-灌-草群落进行合理配植，形成层次丰富、形态优美、极具观赏性的植物景观。

Located by the Lushan Avenue of Shuangliu County in Chengdu, the project covers an area of 4 000m² with completed supporting facilities and amenities, convenient transportation, and pleasant natural environment. The PUD (Planned Unit Development) mode was employed in the project, combing with the introduction of the whole set of classic North American and European housing forms. All residences were integrated, in a low density, into the existing terrain and landscape of the Lushan Mountain.

The most distinctive characteristic of this project is the elaborate design of green space. A beautiful, livable and ecological landscape is created. By zoning different housing areas and emphasizing the spatial layers and orders, the designers attempted to form a series of spaces, from completely open space, to semi-open places, and to private yards that allows residents to experience the changing landscapes. Ecological planting design approaches were adopted in this project. Plant species were selected and combined, and diverse tree-shrub-grass communities formed.

3. 视野开阔的节点空间
4. 花境
5. 水钵
6. 开阔草坪

3. Node space with wide view
4. Flower border
5. Fountain
6. Open lawn

7. 层次鲜明的园路空间
8. 水岸对面的别墅
9. 层次丰富的景观空间

7. Flower border
8. Villas across the lake
9. Series landscape spaces

项目名称 _ 成都万华·麓山国际社区之麓镇样板区
委托单位 _ 成都万华房地产开发有限公司
设计内容 _ 方案及施工图设计
项目面积 _ 4 000m²
设计时间 _ 2011年

Name _ Lu Town Demonstration Area of the Wanhua Lushan International Neighborhood, Chengdu.
Client _ Chengdu Wanhua Real Estate Co., Ltd.
Service _ Landscape design and construction drawing
Size _ 4 000 m²
Year of design _ 2011

人居　Residential Landscape

绵阳长虹国际城
MIANYANG CHANGHONG INTERNATIONAL NEIGHBORHOOD

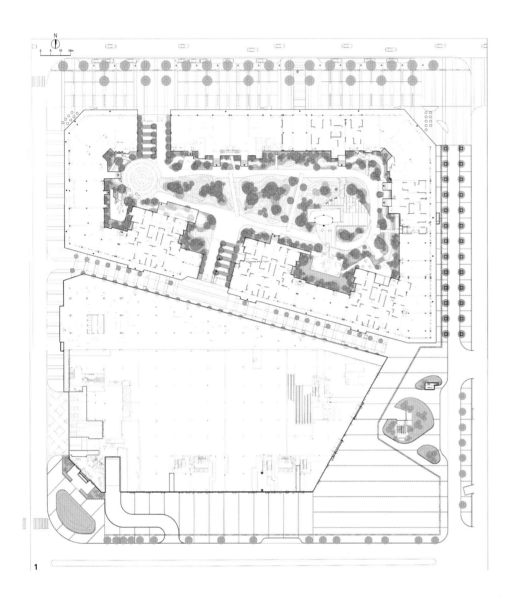

1

长虹国际城作为长虹置业倡导的城市复兴主义的代表作品，由位于北侧的住宅区域及位于南侧的商业区域组成。

住宅区域以"Block"（区块）小空间为规划布局特色，通过建筑围合形成内向型的住宅庭院。采用串珠式的规划布局方式，通过集中绿化面积，减少硬景面积，以达到节约工程造价的目标。在景观用地受限的情况下，设计注重景观的基本功能设置，布置集多种功能于一体的活动场所，利用架空层增加活动空间，并通过道路串联各个功能空间与景观节点。

商业区域的景观设计以打造绵阳"城市客厅"作为其定位，着力塑造"多功能、高品质"的城市空间，通过商业景观设计，满足购物、娱乐、休闲、居住等复合型需求，包含305街区、商业外街、休闲广场、科技文化轴以及屋顶花园五大部分。

（1）老歌新唱的305街区

305街区紧邻跃进路和富乐路，作为原305厂（长虹机器厂）的旧址，这里不仅是一条商业休闲步道，更多的是承载着绵阳人对长虹过往记忆的载体，因此景观设计中除了满足具体

的商业购物需求外，更希望从文化和历史的角度诠释这条步行街的独特性。雕塑、灯具和地面铺装等细节的设计，在满足功能的同时，提供了可供追忆的历史印迹，并提升了购物街区的文化氛围和生活品质。

（2）老少咸宜的商业外街

商业外街保留了原场地的19株大树。街道转角处的绿化隔离带形成半围合的开放空间，结合了入户、人行与车行的三方需求。通过具有历史文化特色的景观小品，来叙述这个城市和长虹的过往与今昔，唤起绵阳老一代人对历史的记忆，同时让年轻人感受到现代与历史相互交融带来的震撼，也让更多的外地来客能在这条街道上看到这座城市历史和文化的缩影。

1. 总平面图
2. 儿童活动空间鸟瞰
3. 通过楼梯来处理场地高差

1. Site plan
2. Bird's eye view of the children playground
3. Stairs are used to solve the height difference problem.

As a representative work embodying the idea of urban renaissance advocated by the Changhong Group, the project consists of the residential area in the north and the commercial area in the south.

The residential buildings were arranged as small blocks, forming many inward courtyards. A necklace-shape layout was adopted to concentrate green spaces in order to reduce the cost on hard landscaping. Considering the limits of the site and the multiple demands of residents, an elevated floor was created for public uses, connecting different service zones and landscape nodes through winding paths.

The design aimed to create a "living room" for the city by building a vibrant, quality commercial complex that can meet diverse needs including shopping, recreation, leisure, and living.

The commercial complex consists of five areas: No. 305 Block, Outer Business Street, Recreational Plaza, Science and Cultural Axis, and roof gardens.

(1) No. 305 Block

As the former site of the No. 305 Factory (also known as Changhong Machinery Factory) which adjacent to the Yuejin Road and Fule Road, the No. 305 Block now is not only a commercial and leisure center of the neighborhood, but also a place that carries the local memories of the Changhong Group. The landscape design needs to respond and reinterpret the glorious history in a modern way. Sculptures, lampposts, and pavements were carefully designed to create a culturally quality atmosphere for the commercial block.

(2) Outer Business Street

Nineteen existing large trees have been preserved in the site. A green belt along the turning corner forms a semi-closed space where not only acts as a main entrance of the residential area, but also separates the pedestrian circulations with vehicle traffics. Landscape features with historical and cultural characteristics are set to narrate the past and the present of both the city and Changhong Group, arouse the local memories, and create an image that links the modern times with the history.

项目名称 _ 绵阳长虹国际城
委托单位 _ 四川长虹电器股份有限公司
设计内容 _ 方案及施工图设计
项目面积 _ 44 147m²
设计时间 _ 2010~2011年

Name _ Mianyang Changhong International Neighborhood
Client _ Sichuan Changhong Electric Co., Ltd.
Service _ Landscape design and construction drawing
Size _ 44 147 m²
Year of design _ 2010 ~ 2011

4. 入口景观
5. 商业区域景观

4. Entrance square
5. Commercial complex

人居　Residential Landscape

成都合能·四季城
HENENG SEASONS NEIGHBORHOOD IN CHENGDU

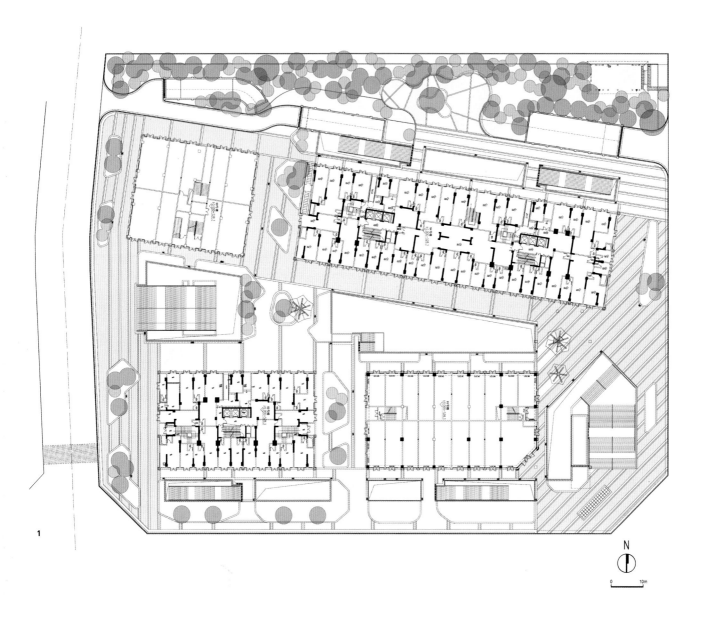

1

合能·四季城位于成都郫县新城——犀浦城西的核心区域，项目占地13 713m²，其中中心商业区占地6 000m²。该项目定位为开放式住宅小区，并在社区中引入了商业街区。

项目设计以鲜明色彩和序列感为主要特色，打造出高品质的商业氛围。项目同时设有由景观廊架、灯具、植物等元素构成的特色休闲场所，及童趣盎然的儿童主题活动空间。

Heneng Seasons Neighborhood is located at the western core area of Xipu Town, a new district of Pixian County in Chengdu, covering an area of 13 713 m², including 6 000 m² for business use. The neighborhood consists of a non-gated residential area and a vibrant commercial area.

Striking colors and strong sense of sequence were

1. 总平面图
2. 入口广场

1. Site plan
2. Entrance square

3. 商业休闲空间
4. 休闲空间错落有致
5. 花朵造型凉亭

3. Recreational space in the commercial area
4. Well-arranged recreational spaces
5. Flower-shaped pavilions

emphasized in the design in order to build a distinctive, quality business center for the new district. Meanwhile, various recreational places formed by landscape corridors, lamps, and plants, and children playgrounds were created to invite all-age users to enjoy the beauty and fun of the site.

项目名称 _ 成都合能·四季城
委托单位 _ 合能瑞进房地产开发有限公司
设计内容 _ 方案及施工图设计
项目面积 _ 13 713m²
设计时间 _ 2012年

Name _ Heneng Seasons Neighborhood in Chengdu
Client _ Heneng Ruijin Real Estate Development Co., Ltd.
Service _ Landscape design and construction drawing
Size _ 13 713 m²
Year of design _ 2012

成都合能·珍宝琥珀
HENENG TREASURE AMBER NEIGHBORHOOD IN CHENGDU

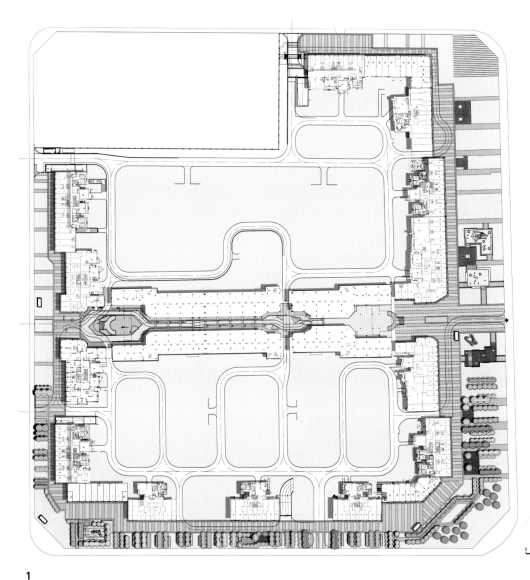

1

  项目位于成都市温江区，由南北两个地块组成，场地四周交通便利。项目总占地约25 800m²（其中包括面积20 000m²的珍宝坊商业街），景观面积达60%以上。

  我们将商业街的使用群体定位为以年轻人为主，提出以"珍宝梦想盒子"作为景观设计主题。通过对周边环境进行调研分析，并对同类型的商业项目进行对比研究，设计师划定了项目中的积极界面与消极界面，并制定出了相应的景观设计策略：根据不同业态功能将积极界面划分成不同的外摆区域，将烘托商业氛围作为设计重点；将富有特色的灯具、景观小品与标识系统等作为消极界面的突破口，力求通过耐人寻味的设计细节来吸聚人气。

  Composed of two neighboring blocks, the project is located in the Wenjiang District of Chengdu City, with convenient transportation infrastructures and supporting amenities. The project covers an area of 25 800 m² (including a 20 000 m² pedestrian commercial street named Treasure Mall), the landscape spaces take up about 16 000 m².

  The commercial street is positioned as a shopping and recreational center which invites young people as its main user group, therefore a theme of "Dream Box" has been proposed. Based on the environmental analysis and the study of similar commercial programs, the designers have developed different strategies to respond active and passive interfaces respectively: the active interface is divided into various outdoor displaying areas to enhance the commercial atmosphere; in the passive interfaces, landscape features such as lamps, street furniture and structures, signage and interpretation systems are employed to attract and gather visitors.

1. 总平面图
2. 入口景观
3. 商业区步行空间

1. Site plan
2. Landscape of the entrance area
3. Pedestrian spaces in the commercial area

项目名称 _ 成都合能·珍宝琥珀
委托单位 _ 合能瑞进房地产开发有限公司
设计内容 _ 方案及施工图设计
项目面积 _ 25 800m²
设计时间 _ 2012年

Name _ Heneng Treasure Amber Neighborhood in Chengdu
Client _ Heneng Ruijin Real Estate Development Co., Ltd.
Service _ Landscape design and construction drawing
Size _ 25 800 m²
Year of design _ 2012

成都汇锦城
CHENGDU HUIJIN COMPLEX

成都汇锦城位于成都市高新南区，项目融合了格调餐饮、时尚生活、风情休闲、浪漫体验四大主题，并将引进星级院线、大型购物卖场、品牌旗舰店、奢侈品馆等设施，力求打造一条高品质的综合商业水景街。

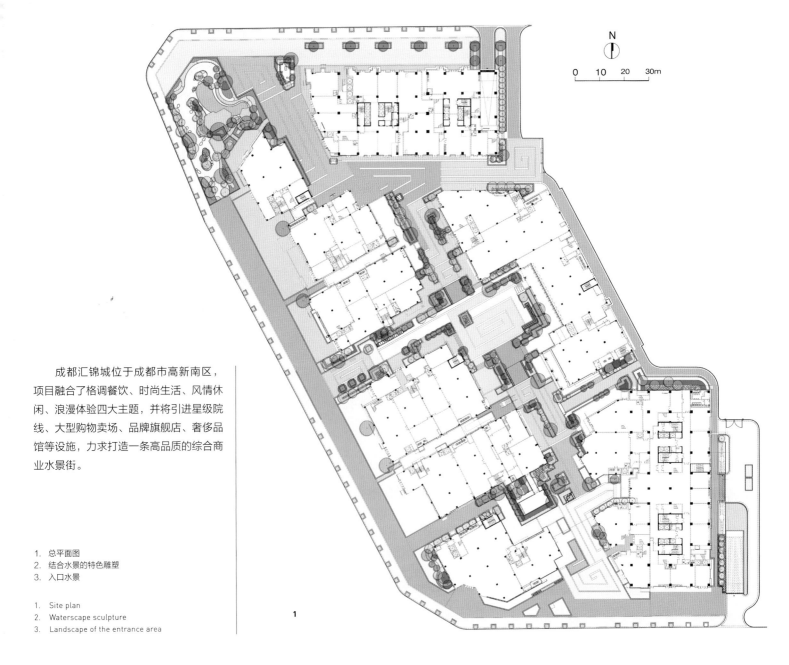

1. 总平面图
2. 结合水景的特色雕塑
3. 入口水景

1. Site plan
2. Waterscape sculpture
3. Landscape of the entrance area

2

3

人居　Residential Landscape

4. 商业街景观鸟瞰
5. 休闲区域

4. Bird's eye view of the commercial street
5. Tranquil leisure space hidden in the street

Huijin Complex is located in the South Gaoxin District of Chengdu, and positioned as an integrated place of restaurants, shops, and recreation and experience spaces which are connected by various landscape water features. Theatres, supermarkets and malls will be introduced into the site in the near future to complete and improve the amenities of the quality commercial street.

项目名称 _ 成都汇锦城
委托单位 _ 成都裕丰汇锦置业有限公司
设计内容 _ 方案及施工图设计
项目面积 _ 14 000m²
设计时间 _ 2014年

Name _ Chengdu Huijin Complex
Client _ Chengdu Yufeng Huijin Real Estate Co., Ltd.
Service _ Landscape design and construction drawing
Size _ 14 000 m²
Year of design _ 2014

成都伊泰天骄

YITAI TIANJIAO NEIGHBORHOOD IN CHENGDU

伊泰天骄项目坐落于成都东二环与东大路交汇处，周边功能设施配套齐全，交通便捷，邻近中、高等院校及三大文化休闲公园，区位资源得天独厚。项目占地约5.4hm²，建筑面积31万m²，由12栋错落有致的高层建筑组成。

伊泰天骄景观设计项目包括售楼部景观改造、住宅园区景观改造及架空层景观设计。设计师将当地文化、项目自身特点纳入考量，为其量身定制"新衣"。设计以项目已建成部分为基础，对其主题进行了重新定位，并对景观总平面图进行重新梳理和规划，最终形成了"千鸟千树"绿影长廊、"十二大亲子社区"主题馆、"蓝色森林"游园区、"樱花海"运动场四大景观亮点，营造出可观、可赏、可游的高品质居住社区。

1. 售楼部
2. 总平面图
3. 入口景观

1. Sales center
2. Site plan
3. Landscape of the entrance area

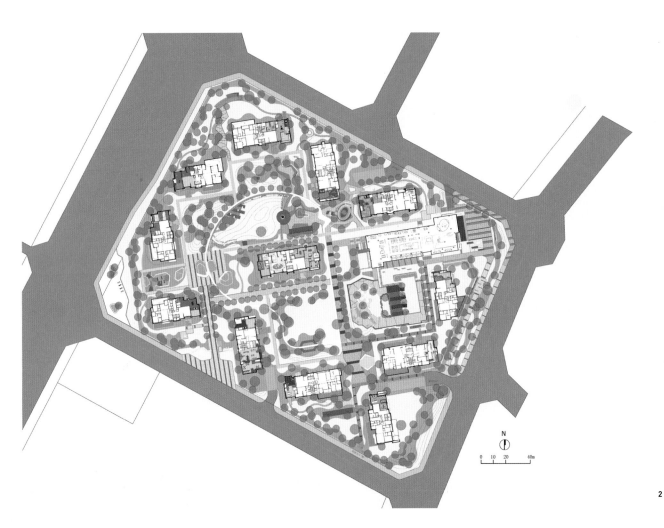

人居　Residential Landscape

Yitai Tianjiao Neighborhood is located in Chengdu, at the intersection of the east second ring road and Dongda Road, with complete infrastructures and civic amenities. The site neighbors several high schools and colleges and three major cultural parks, offering its regional location a great advantage. The project consists of 12 well-arranged high buildings, covering an area of 5.4 hm², and 310 000 m² floor area.

The landscape design includes the renovation design for the sales center, the residential area, and the elevated floor. Taking local cultural elements and the project's unique qualities into consideration, the design team reshapes a new image for the site. Based on its built structures, the design repositions the project by evaluating and re-organizing its existing features and resources, and creates four landscape zones, including a green shadow corridor where people can observe birds and tall trees, a theme pavilion which provides twelve playing areas for children and their parents, a garden named "Blue Forest" and a playground called "Sakura Ocean." All these facilities help to create a high-quality, joyful residential community.

4, 5. 住宅区景观
6. 架空层平面图
7. 架空层海洋馆

4,5. Landscape of the residential area
6. Plan of the elevated floor
7. Aquarium on the elevated floor

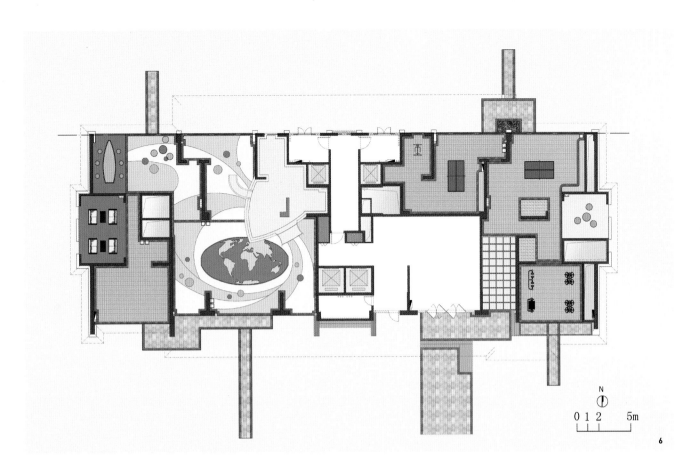

6

7

项目名称 _ 成都伊泰天骄	**Name** _ Yitai Tianjiao Neighborhood in Chengdu
委托单位 _ 伊泰置业（成都）有限公司	**Client** _ Yitai Real Estate Co., Ltd. (Chengdu)
设计内容 _ 方案及施工图设计	**Service** _ Landscape design and construction drawing
项目面积 _ 40 000m²	**Size** _ 40 000 m²
设计时间 _ 2015年	**Year of design** _ 2015

人居　Residential Landscape

致　谢

景虎十年，感恩有您！

"FOR LAND, FOR YOU"，景虎一直铭记于心。

徐徐回望，几多辉煌，几多波折，

感谢心系景虎的你我他！

无论明天怎样变幻，

"执着于景，敬天爱人"的初心不改。

诚邀您一起，

见证景虎的蜕变！

龙赟

2016年3月

于成都

ACKNOWLEDGEMENTS

We appreciate your support and contribution in the ten-year history of LANDHOO!

"FOR LAND, FOR YOU" is always the motto to LANDHOO.

Glories and setbacks interweave,

Thanks to everyone who devoted to LANDHOO!

No matter how things change in the future,

The initial faith of "ardent love for landscape and great respect for nature and human" stays.

We invite you

To follow LANDHOO's growth!

YUN LONG
March 2016
In Chengdu

图书在版编目（CIP）数据

景观森林：设计实践 / 龙赟著. — 南京：东南大学出版社，2016.8

ISBN 978-7-5641-6611-3

Ⅰ. ①景… Ⅱ. ①龙… Ⅲ. ①景观设计 Ⅳ. ① TU986.2

中国版本图书馆 CIP 数据核字（2016）第 154751 号

景观森林设计实践

出版发行：东南大学出版社
出 版 人：江建中
责任编辑：朱震霞
社　　址：南京市四牌楼 2 号　邮编：210096
网　　址：http://www.seupress.com
电子邮箱：press@seupress.com
经　　销：全国各地新华书店
印　　刷：上海利丰雅高印刷有限公司
开　　本：889mm×1194mm　1/12
印　　张：16.5
字　　数：250 千字
版　　次：2016 年 8 月第 1 版
印　　次：2016 年 8 月第 1 次印刷
书　　号：ISBN 978-7-5641-6611-3
定　　价：115.00 元

本社图书若有印装质量问题，请直接与营销部联系。电话：025-83791830